Hollywood Films about Schools

Hollywood Films about Schools: Where Race, Politics, and Education Intersect

Ronald E. Chennault

palgrave
macmillan

HOLLYWOOD FILMS ABOUT SCHOOLS

First published in 2006 by
PALGRAVE MACMILLAN™
175 Fifth Avenue, New York, N.Y. 10010 and
Houndmills, Basingstoke, Hampshire, England RG21 6XS
Companies and representatives throughout the world.

PALGRAVE MACMILLAN is the global academic imprint of the Palgrave Macmillan division of St. Martin's Press, LLC and of Palgrave Macmillan Ltd. Macmillan® is a registered trademark in the United States, United Kingdom and other countries. Palgrave is a registered trademark in the European Union and other countries.

ISBN 1–4039–7293–1

Library of Congress Cataloging-in-Publication Data

Chennault, Ronald E.
Hollywood films about schools : where race, politics, and education intersect / Ronald E. Chennault.
p. cm.
Includes bibliographical references and index.
ISBN 1–4039–7293–1
1. Schools in motion pictures. 2. Education in motion pictures. 3. Race relations in motion pictures. 4. Motion pictures—United States. I. Title.

PN1995.9.S253C44 2006
791.43′6557—dc22 2005056646

A catalogue record for this book is available from the British Library.

Design by Newgen Imaging Systems (P) Ltd., Chennai, India.

First edition: March 2006

10 9 8 7 6 5 4 3 2 1

Printed in the United States of America.

Contents

Acknowledgments

Let me begin by thanking Amanda Johnson at Palgrave for the opportunity to publish my work. Airié Stuart, Erin Ivy, and Petrina Crawford deserve my gratitude as well.

I am also indebted to Aaron Gresson III, Henry Giroux, Joe Kincheloe, and Patrick Shannon for the important roles they all played in shaping this book in its earlier form. My special appreciation goes to Aaron for the brotherhood and mentorship from which I have benefited over the years.

Several other people need to be acknowledged for their particular contributions, large and small. Nell Cobb, Layla Suleiman, and Sheldon Woods created a circle of sanity when I needed it most and helped me to work through some difficult professional decisions. Leila Villaverde has been perpetually willing, even across the miles, to offer her assistance whether I took her up on the offers or not. James Earl Davis has listened more than most, sharing in my frustrations and celebrating my good news.

I owe a special thanks to Derrick Alridge, who is responsible in more ways than one for helping to see this project to its completion.

Finally, my deepest thanks go to my mother, father, and brother for their unconditional love and support. I have been truly blessed by them and through them.

1

Introduction

The film *Dangerous Minds* (1995) is an indisputable product of the Hollywood formula. It is, as David Bordwell might put it, so "obvious" an example of classical Hollywood cinema that it almost goes without saying (Bordwell et al. 1985, 3). In fact, the film is so typical in many respects that it does little to "defamiliarize" itself—which some (e.g., neoformalists) believe is the function of art—except with regard to its cast of characters.[1] But to describe the film as typical does not mean that it is uninteresting. One interesting aspect of the film, for instance, can be found in its ideologies, which might be defined as the values and goals that constitute its agenda. Ideologies are used to make sense of society and to regulate social practices (van Dijk 1998).[2] On the one hand, the principal ideologies of the movie as I perceive them—the reassertion of patriarchy and white racial superiority—are not especially unusual in Hollywood cinema. Yet the film's attempt to hide its politics underneath a feel-good Disney narrative, especially within the then-existing sociopolitical climate in the United States, is the fascinating aspect to which my attention is drawn.

Watching the film for the first time prompted me to think about its underlying messages in relation to other films I had seen and caused me to wonder if the politics and representations in *Dangerous Minds* had been present in other texts. The questions that resulted from my reflections and my discussions with others on this subject were even more specific: Are these representations (especially as they concern race) common among Hollywood films? Among Disney films? Among films in general? Is the appearance of these elements a recent occurrence? Have they always been present? In what forms?

As an educator who is concerned with the pedagogies that are employed both in schools and outside of them, I was also intrigued that these ideologies were being promoted within the context of a film primarily about education. Thus, I became more committed to examining what *Dangerous Minds* had to say about race, politics, and education. In light of the film's plot, ideologies, and racialized discourse, and given my keen interest in these elements (both in this particular text and in the main), I expanded my musings into a larger project, the result of which is the book. My aims were to analyze the representations of race in Hollywood films about educational institutions ("school films"), as well as to explore how these films might be used pedagogically. Most of the films foreground the issue of race, so it is an especially suitable topic of study in these texts. As for the term "pedagogy," I use it in this context to mean, generally speaking, what educators can teach about these films and how they (we) might carry this out.

In his posthumously published book, *White Screens, Black Images*, James Snead (1994) raises a rarely articulated issue concerning the difficulty of studying visual texts: "Since many mass-media images today claim to be neither reality nor fantasy (witness the docu-drama), there are no useful criteria by which to inspect or challenge the claims to truth that these visual images and events constantly make" (141). His point is well taken. Without calling for textual "realism," however, one can and must hold these visual texts responsible for their "claims to truth," and acknowledge that there are criteria for judging how responsible these texts (and their producers) are. But how should one attempt to inspect these claims to truth in a systematic fashion?

This is a question with which I wrestled upon beginning my study. To address this dilemma, I chose to bring to bear the insights and analytical tools of critical pedagogy and media and cultural studies on a specific set of Hollywood films dealing with schooling and educational institutions.

This group of films, all feature-length, form what some might consider a subgenre of considerable size, extending from *Goodbye, Mr. Chips* (1939) to present-day cinema.[3] However, there are a few major factors that I used to select films from among this subgenre of school films. The first was that they be centrally concerned with some act of educational intervention, carried out either by a classroom teacher or by a school principal, within the context of a school. This criterion, then, excludes a large number of Hollywood movies for which the school campus merely serves as a backdrop for other plot concerns, such as adolescent development (e.g., *Fast Times at Ridgemont High* (1982)) or athletic competition (*The Program* (1993)).

Another main criterion for choosing films involved in the study was that each film was made between 1980 and 2000, when an abundance of school films appeared on the silver screen. Staying within this time period enabled me to make some connections to the social, political, and cultural conditions that existed during and after the presidency of Ronald Reagan. By now we have some critical distance from that time period, which allowed for a more informed examination of the political and cultural climates during those two decades. To be sure, there are some noteworthy commonalities between the school films I studied and those that were released before this period, and I make reference to these earlier films when pertinent. My main focus, though, centers on the smaller number of school films made in the 1980s and 1990s, with the hope that the analysis that I provide will have some applicability to a broader range of filmic texts.

Moreover, the films in my sample either were rather large box-office draws during their respective theatrical releases or received a significant amount of critical attention, or both. This, of course, excluded from consideration most lesser-known or independently distributed films. This additional factor allowed me to focus on films that were most popular or that fit securely within the mainstream, which should enable my sample of films to be more fairly compared with the racial representations of other mainstream Hollywood movies.

The Framework

Because of the intricacy of the confluence of racial representations and politics that exists in the films under examination, I found it beneficial to draw upon the perspectives of a number of fields of study instead of searching for one "grand theory" or guiding discipline. Such an approach can be considered interdisciplinary or "theoretical bricolage" (Denzin and Lincoln 2001, 6). What these terms emphasize is the usefulness of building a framework based upon the insights of more than one perspective or field. Drawing on more than one theoretical body of work is not an effort to escape scholarly soundness; it is instead meant to yield a more sophisticated analysis of the subject of study and to avoid falling into the trap of "theoretical absolutism" (Johnson 1980, 5; Beezer et al. 1986; Thompson 1988).

One of the perspectives that informs my analysis comes from the field of critical pedagogy. One model of critical pedagogy that I draw upon evolved during the 1980s as a result of the influence of the critical theory associated with the Marxian Frankfurt School, hence the modifier "critical." Although the scope of critical pedagogy has changed over time, there are certain issues that are within the purview of most approaches to the field. Giroux (1994) has indicated some of these issues or "problems" with which critical pedagogy has been concerned: "the relationships between knowledge and power, language and experience, ethics and authority, student agency and transformative politics, and teacher location and student formations" (131). Elsewhere, Giroux, Simon, et. al. (1989) point to the primary value of critical pedagogy for my project; that is, its commitment "to read the ground of the popular for investments that distort or constrict human potentialities and those that give voice to unrealized possibilities" (25).

My interest in critical pedagogy is also linked to a desire to see knowledge contextualized and related to people's lived realities. In cultural studies—with its concern over the production of knowledge, values, and identities within historical, social, political, institutional, and cultural sites—lies fertile ground for the fulfillment of this desire. Stuart Hall has claimed that cultural studies "insists on the necessity to address the central, urgent and disturbing questions of a society and a culture in the most rigorous intellectual way we have available" (1992, 11). This underscores the need for the mutual dependence of cultural studies and education (Johnson 1980), through which theory and practice are combined in order to negotiate the connections between culture, power, and agency.

More specifically, Giroux (1994) has discussed at length the importance of integrating critical pedagogy and cultural studies in the analysis of popular cultural texts.[4] Related to Giroux's call for the intersection of critical pedagogy and cultural studies is Kellner's (1995) attempt to define a cultural studies that is "critical, multicultural, and multiperspectival." For example, one of Kellner's criteria for criticalness is that cultural studies be "interested in advancing a critique of structures and practices of domination and advancing forces of resistance struggling for a more democratic and egalitarian society" (95). Additionally, cultural studies should stress commonalities as well as differences, and not just difference for the sake of difference, but differences that matter.

This, of course, is not a simple exercise. Attempting to include serious discussion of popular culture in educational discourses, for instance, has become much more commonplace in the past decade. Yet it can still be a controversial and at times highly contested act to the degree that it "presents an urgent, fundamental, and undeniable challenge to pedagogical theory and practice" (Collins 1994, 56). Ava Collins makes the following insightful observations about responses from the Right and from the Left to popular culture:

> [T]he Right wants to "reclaim the legacy" [in William Bennett's words] . . . , and in effect banish popular cultural forms from the classroom. The Left is struggling to develop analytical models that take into account changed social, cultural, and material conditions, but ironically they often fail to extend their analyses to the very system within which they are developing those models, thus frequently reproducing the very relations of cultural and social power that they seek to

> investigate, rarely questioning their own positions within an educational system that produces, organizes, and legitimates certain means or methods of the transmission of knowledge and power. (1994, 56)

Hence we can see how exploring popular culture as a pedagogical site can pose a challenge to educators of all stripes. This is a challenge, however, that needs to be and is being faced head on by various educators.

For instance, recognition and exploration of some of these cultural texts as sites of knowledge production takes place in media literacy, a field that has much in common with critical pedagogy and cultural studies. David Trend argues that media literacy's approach

> entails abandoning assumptions that particular readings are self-evident or that the medium itself is a neutral carrier of information . . . In contrast to these views, it needs to be pointed out that truth does not pass perfectly through a video to a student, nor knowledge through a teacher. Both require the engagement of students in receiving the messages and making sense of them. (1994, 238)

The idea that truth does not reside in visual texts may be a trite declaration to some, but given the widespread adherence to the adage, "Seeing is believing," it seems that we need to continue to debunk this myth. The unreliability of this hoary maxim is evidenced by the multiple, competing ways in which people "saw" what happened to Rodney King upon viewing the infamous videotape of his beating. David Buckingham underscores the pedagogical importance of reconceptualizing how we study the media when he contends that "the argument for Media Studies is not merely for changing the content of the curriculum: it is also implicitly an argument for developing new approaches to teaching and learning" (1991, 33).

Integrating the theoretical perspectives of critical pedagogy, cultural studies, and media literacy undoubtedly ruptures some traditional analytical practices, but this break with tradition has the potential to be more productive than destructive. My examination of the intersection of race, politics, and pedagogy in Hollywood school films could not help but benefit from the multiplicity of approaches made possible by these perspectives. Their concern with examining the interplay between knowledge and power, with transcending the rigid distinctions between academic disciplines in order to (re)contextualize knowledge, with exploring the significance of textual representations in our lives, and with comprehending the workings of media, culture, and society are what necessitated the inclusion of these perspectives in the study.

The Analysis

Hall (1981) argues that, "amongst other kinds of ideological labour, the media construct for us a definition of what *race* is, what meaning the imagery of race carries, and what the 'problem of race' is understood to be. They help to classify out the world in terms of the categories of race" (35). In light of Hall's cogent argument, my examination explores the stances my selected films take on the subject of race as exhibited by the ideologies they rely on and the lessons they attempt to teach. This process mainly involved conducting close readings of the following films: *Class of 1984* (1982) *Teachers* (1984), *The Principal* (1989), *Lean on Me* (1989), *Dead Poets Society* (1989), *Renaissance Man* (1994), *Dangerous Minds* (1995), and *Music of the Heart* (1999). More specifically, I studied how racial subjects are positioned within the texts, how power is distributed among characters of different races, and how the characters address, avoid, or ignore racial difference. In this way, I tried to discover and address what these films "say" about race.

In his survey of research on knowledge construction via popular culture, Cortés (2004) identifies four dimensions of analysis among this literature: content, control (or production), impact (or effects), and pedagogy. In a different but related context, Kellner (1994) argues that the field of cultural studies, should, at its best, "discuss production and political economy, engage in textual analysis, and study the reception and use of cultural texts" (105). The book that follows does not fully explore each of the analytical dimensions nor is it as comprehensive in its approach as Kellner recommends. However, this degree of comprehensiveness is a worthwhile goal (especially for the entire field of cultural studies), and I endeavored to honor it in my work. To do this, I analyzed the content of the texts, aspects of their production, a few indicators of the reception of these texts, and, lastly, their pedagogical implications. In order to enhance this analysis, it was necessary to locate the chosen films within a social and historical context. Ultimately, I attempted to make connections between the ideologies present in the films and in society at large, specifically as they concern race. My goal was not to make some simplistic pronouncements about how race is represented in these films (such as, "This film is racist."), but to take note of how the films rely upon and contribute to existing racialized discourses. I do expect, however, that my work will help to underscore the absurdity of Dinesh D'Souza's (1995) contention that racism has "ended."

The complex way in which race is inextricably linked to American culture guarantees, however, that this will not be an easy or straightforward task. This task is further complicated by the problematic nature of the category of "race" itself. I acknowledge that race is not a fixed, immutable, biologically determined trait or set of traits (although race has been invested with certain physiognomic dimensions). Rather, the meaning of race has emerged from particular social, historical, political, economic, cultural, and even religious conditions. But to deny the existence of an essentialized racial subject is not to deny the power and importance that the category of race carries. In this regard, Omi and Winant's (1986) conception of race is instructive, which they explain thusly: "[A] concept which signifies and symbolizes social conflicts and interests by referring to different types of human bodies" (54).

For the purposes of this study, I considered all of the characters in the films to be racialized; moreover, I tried to discern and comment about the racial discourses and ideologies that are implicated in the films. The primary (but hardly the only) terrains that were searched for these ideologies were the discourse and the practices of the central pedagogical figure. This strategy did not, of course, preclude considering the actions of the other characters in the films. Indeed, it was necessary to consider all of the characters as well as each film's aesthetic and political dimensions in order to achieve a more complete understanding of what is at work. The assumption behind my strategy is merely that the main school authority figure is highly representative of the films' politics since the selected films build their respective stories around these figures.

The chapters that follow feature an analysis of the political, cultural, and educational implications of the racial representations of the school films from the 1980s and 1990s. I begin in chapter 2 with a review of the platforms and policies of the Reagan administration, particularly with regard to the administration's education and civil rights agendas. I consider as well the mood of the country under Reagan's leadership and the socioeconomic impact of his legacy. Chapters 3, 4, and 5 consist mainly of close readings of the films in my sample, which I situate within the concurrent social and political climate in America. These chapters are organized in thematic fashion, generally corresponding to changes in the Oval Office. Throughout these chapters, however, is a consistent emphasis on the influence of the Reagan legacy in the narratives of the films. The last chapter deals with how the school movies and my analysis of them might be used pedagogically. On the whole, I do not pretend that the study is all-encompassing, but I do see it as a detailed and extensive examination that may be of benefit to a variety of audiences.

2

Reagan, 'Rithmetic, and Race

Ronald Reagan often seemed to perceive the American presidency as yet another acting job, probably the most significant role of his life; therefore, it seems a good place to begin this study with an examination of the interrelatedness of the dominant political agenda, the country's cultural environment, and Hollywood cinema during the Reagan era. In particular, this chapter is concerned with the events and episodes of the Reagan era that contributed to and benefited from the turbulent political atmosphere out of which the school films in this study emerged. There are innumerable aspects on which I could concentrate, but I focus narrowly on the ones that are most relevant to the interests of the study. The chosen aspects are the educational agenda of the Reagan administration; its civil rights record and broader racial politics; the conservative environment in the United States during Reagan's presidency; and the display of a corresponding conservatism in American cultural expressions, especially Hollywood movies. My discussion of these few but important elements should underpin the analyses of the school films that appear later in the study.

Reagan and Educational Policy

It is Steven Shull's contention—a primary theme of his 1993 text, in fact—that "presidential influence leads to policy change." With regard to former presidents Reagan and Bush, this influence is particularly evident in the changes in the national education agenda. The Department of Education (ED), which received a separate cabinet-level position near the end of Jimmy Carter's presidency in May of 1980, was seen by Reagan as an unimportant and even unnecessary department. He perceived its existence as a giant step toward a nationalized school system (Dugger 1983). In fact, the Secretary of ED post was the last to be filled by Reagan after his election, and the entire department was slated to be abolished and replaced with "a newly established agency that would be less powerful and prominent in the total structure of the federal government" (Bell 1986, 488).

In her groundbreaking study of conservative ideology, federal education policy, and public schooling during Reagan's first term, Catherine Lugg (1996) concludes that the former president's approach to education was informed most principally by social traditionalism. She documents the Reagan administration's repeated attempts to build political appeal for school prayer, tuition tax credits, and vouchers. The administration toiled to give the impression of being sincerely concerned for what was missing in education, and "it unintentionally ignited a school reform movement" by co-opting *A Nation at Risk* for its own political gain (Lugg 1996, 205). Yet, in apparent contradiction to the call in *A Nation at Risk* for the federal government to help fund and support the national interest in education, Reagan asked Congress to merge federal funding for disadvantaged, physically challenged, and bilingual children into two block grants and reduce the money by 25 percent and further requested that overall federal aid to education be sliced in half by 1986 (Dugger 1983, 319).

Reagan was ultimately unable to disband the department during his two terms in office, but he appointed three secretaries who were to varying degrees responsible for reducing educational support at the federal level. The first secretary, Terrel H. Bell, fought to keep ED intact in spite of Reagan's wishes (and the wishes of other influential members of the administration) to the contrary. Bell, while largely praiseworthy of Reagan's leadership and conviction, has been indirectly critical of the administration's lack of a formal policy statement regarding education during his four-year tenure (Bell 1986; Lugg 1996; Stallings 2002). The second appointee, William J. Bennett, who served as secretary for most of Reagan's second term, was the most familiar, most outspoken, and most conservative of the three. Bennett was and is characteristically Republican in his calls for less government support and more individual responsibility for educational funding. Often, however, his attention veered away from the educational concerns of the country and focused on attacking drug use, an attack with racial undertones frequently just below the surface.[1] His was an easy transition, then, to the "Drug Czar" position created by Congress in 1988. Last of the three was Lauro F. Cavazos—the first Hispanic cabinet member—who was an eleventh-hour Reagan appointee from Texas. To the extent that his politics were known prior to his appointment, he was seen as much more moderate than the outgoing Bennett (although views later expressed by Cavazos converged somewhat with those of Bennett and Reagan (Cavazos 1990)). For this and other reasons, Reagan's choice of Cavazos was believed to have been simply a means of attracting Hispanic voters for then-Vice President Bush's future presidential race (Miller 1988).

As one might expect, the harshest and most regressive policies came under the first two secretaries. For example, while state and local funding increased dramatically, the federal contribution to education fell from 9.2 percent to 6.2 percent of the total national education budget from 1980 to 1987. Programs such as Chapter 1 (which was reauthorized under its original name of Title I in 1994) and Head Start, as well as higher education student aid programs—from which nonwhite and poorer students disproportionately benefited—experienced severe cutbacks (Simmons 1984; Hylton 1988).

Generally, the Reagan administration had little regard for public education in America—that is, until the National Commission on Excellence in Education released its report, *A Nation at Risk*. After the commission study had been heavily reported on in the press, Reagan started speaking publicly on educational issues at every turn. The dramatic portrayal of the state of American education sketched out in the report appealed to Reagan; it featured the kind of "apocalyptic" language that he and his administration preferred in other areas of analysis.[2] According to Bell, although Reagan wanted to reduce federal spending for education, he did not advocate abandoning federal financial assistance altogether. "To understand the lack of a consistent, rational Administration policy with respect to education," Bell writes, "it is necessary first to understand the agenda of the movement conservatives both within and outside the Reagan Administration. These people looked on the election of Ronald Reagan in November 1980 as the beginning of a revolution" (Bell 1986, 490). Whether the cause was this "radical conservative movement" or Reagan's own directives, the Department of Education under Reagan evolved into a conservative stronghold intent on promoting an educational agenda largely ruled by free-market sensibilities.

The Racial Politics of the Reagan Administration

Reagan began a three-day leg of his 1980 presidential campaign by appearing at a county fair in Philadelphia, Mississippi. During his speech, he told the audience that he would work "to restore to states and local governments the power that properly belongs to them" (Cannon 1980). Despite Reagan's claim of not remembering this event (Rowan 1991), he and his advisers clearly understood at that time the potential significance of his decision to appear there and to employ the language of "states' rights," especially given Reagan's keen appreciation of the role of symbolism and rhetoric. After all, this was the site of the horrific slayings (involving a sheriff, local police officers, and the Ku Klux Klan), of three young, male, civil rights activists—one black, two white—who had traveled there in the summer of 1964 to register black voters. By the time of Reagan's visit, no one had been convicted of the murders.[3] Furthermore, states' rights had been closely associated in the 1950s and 1960s in the South with support for segregation and Jim Crow laws. Regardless of the motivations behind Reagan's choices, they served as a harbinger of the shifts in racial politics and in the civil rights agenda that would transpire under the Reagan administration.

The conservatism of Reagan's educational agenda discussed above was matched by the initiatives of his administration in the civil rights arena. In fact, his civil rights agenda is better described as activist, given that its driving force was the intentional reversal of that which had been expanded under previous administrations. During the 1980s, some racial minorities experienced a few symbolic and material gains, such as high-ranking appointments in the federal government and measurable job growth. However, the setbacks caused by the administration far outweighed those gains or often placed them in jeopardy. For one thing, Reagan did not make his appointments to important positions of civil rights leadership with the best interests of nonwhites in mind.[4] Second, under the Reagan administration, support offered to racial minorities who were discriminated against in educational institutions, in funding for public education, and in federal efforts to enforce desegregation policies all declined. In addition, nonwhite citizens—those of lower socioeconomic status in particular—also suffered from a reduction in employment opportunities and job training programs. Worse yet, Reagan and his supporters waged a relentless attack on racial minorities by attempting to reverse affirmative action rulings, attempts that have continued to this day. Behind this agenda, Crenshaw contends, "was a formalistic, color-blind view of civil rights that had developed in the neoconservative 'think tanks' during the 1970's" (1988, 1337).

One way President Reagan dealt a crippling blow to the civil rights struggle was through his federal appointments. He used his power to weaken the strength of the civil rights policies that prevailed before his presidency, many of which had been established under Lyndon Johnson and had just recently gained support from Jimmy Carter (Shull 1993). Reagan claimed to be committed to appointing more racial minorities and women to top judicial and policy posts than any previous administration, but his claim is quite misleading. The U.S. Commission on Civil Rights (CRC) performed a study that revealed that the number of minority and female appointees declined proportionately during Reagan's first term in office in 33 of the categories studied, and had only risen in 8 categories (Miller 1984). Moreover, Reagan made 294 appointments to 752 federal judicial posts in his first six years in office, and 366 appointments in all, the majority of whom were young, white, male lawyers who belonged to the ideological Right (Ruffin 1987; Guinier 1989).

Reagan's appointment of conservative figures who were not recognized civil rights activists to positions of civil rights advocacy was problematic as well. William Bradford Reynolds, for example, was appointed to head the civil rights division of the Department of Justice (DOJ). With no experience in or known commitment to civil rights activism, Reynolds could hardly have been expected to help promote the rights of victims of racial discrimination. Unsurprisingly, he propagated the conservatism of the administration through his immediate efforts to curtail school busing and affirmative action programs. Other such appointments soon followed, such as Charles Cooper (former law clerk to conservative Chief Justice William Rehnquist) who became deputy to Reynolds, and William French Smith to the office of Attorney General.[5]

Probably the most controversial appointive measure of the Reagan administration, however, was the restructuring of the CRC through the replacement of existing members with anti–civil rights personnel. These replacements were controversial for two important reasons: on the one hand, the traditionally independent and aggressive CRC was being transformed into a den of conservatism; on the other, it was the first time any president had removed incumbent members of the CRC. Althea Simmons (1984) argues that "[t]he Commission, which had criticized every previous administration when its policies did not serve to advance civil rights, [was] effectively muzzled as an advocate for civil rights" and racial equality (9). Although Reagan received vehement criticism about his choices to the CRC and to other civil rights agencies, he made no effort to rescind any of his decisions. The actions of the Reagan administration and the haste in which they were enacted—the aforementioned decisions were made within the first few months of Reagan's term—are clearly indicative of the administration's intentions of mitigating the strength of the civil rights movement.

The educational agenda during the Reagan era also proved to be regressive in terms of racial equality. Under Attorney General Smith, the DOJ sought to limit Supreme Court jurisdiction and challenged the long-standing ban issued by the Internal Revenue Service on tax exemptions for racially discriminatory private schools. The administration attempted to persuade the Supreme Court that the IRS ban was illegal because the agency had no right to enforce social policy (Shapiro and Camper 1983; Miller 1984; Shull 1989).

Evidence of Reagan's detrimental impact on racial minorities can be found on a larger scale in his administration's policies concerning public school integration. When the Reagan administration assumed control of the DOJ, much progress had been made in the struggle for racially integrated schools. Yet this did not deter the administration from challenging the school desegregation law (Sussman 1984). William Reynolds stated before the House Judiciary Committee that he was not in favor of involuntary school busing.[6] He made it clear that racial integration in the schools should only occur on a voluntary basis, even though past voluntary desegregation efforts had not succeeded. To support his position, Reynolds employed many arguments that had been rejected by the Supreme Court many years earlier. Neely (1984) has argued that under Reagan and his DOJ—and under Reynolds in particular—the administration was less aggressive in identifying discrimination against nonwhites and women and tried to narrow the avenues available to correct discriminatory practices (1772).

However, the administration was not consistent in its support of voluntary racial integration. For example, local school authorities in Seattle, Washington, voluntarily adopted a school desegregation plan. But the plan was rejected by voters in a statewide election. The Seattle Board of Education in turn argued that the state referendum was unconstitutional and was motivated by racial prejudice. Despite the Reagan administration's claims that it advocated voluntary desegregation and control of education at the local level, it supported the state's referendum. Once again, the Supreme Court intervened and ruled against the state and the DOJ (Days 1984; Sussman 1984). These actions regarding school desegregation exemplify how the administration "inadequately enforced and otherwise undermined, if not violated outright, settled law in the field of civil rights" (Days 1984).[7]

Some Reagan supporters have argued that racial minorities, specifically blacks, have benefited educationally from the actions of the Reagan administration. In his article on the prosperity of blacks in the 1980s, Joseph Perkins (1988) asserts that there was an increase in black college enrollment and in the black high school graduation rate, and that Reagan should be given credit for fostering the economic climate that produced these increases. Perkins provides some data to support his assertion, such as the contention that "even though the number of college-aged blacks has decreased this decade, black college enrollment increased by 4 percent between 1980 and 1985, to 1.05 million from 1.01 million" (28). Although the data may be accurate, it is at best speculative to conclude that the growth in enrollment was occasioned by the Reagan administration. In fact, it is equally plausible to argue that this increase, coming in the years after the Carter administration, could be attributed to strides made toward racial equality under his leadership; or, that this increase could have been spawned by the civil rights activism of the preceding decade.

There are some results, however, that were much more obviously brought about by the administration. For instance, black students comprised 35 percent of those affected by Chapter 1 education aid for disadvantaged elementary and secondary school students. Spending for Chapter 1 under Reagan declined by 19 percent. Even more decreases were proposed by the administration, but were disallowed by Congress. Also, black students account for one-third to one-half of the participants in higher education student aid programs. Funding for one such program, Guaranteed Student

Loans, was reduced by 26 percent. Thus, these significant program cuts unavoidably affected blacks and other racial minorities since they make up a large part of the aid recipients (Simmons 1984).[8]

The effect of the Reagan administration on employment opportunities for blacks is still another oft-debated issue. One argument is that, as a result of Reagan policymaking, blacks saw impressive growth in jobs in the 1980s. Perkins (1988), for example, states that the rise in black incomes is a result of job growth over the five-and-a-half year period from mid-1982 to 1988. He claims that total black employment during this time grew by about 2.3 million people; that the black unemployment rate has been cut by almost half since 1982, and the employment rate increased from 49 percent in 1982 to 56 percent in 1988; and that from 1983 to 1988, 122,000 black teenagers gained employment, causing unemployment for black teens to be lowered by 20 percent over the period. Additionally, he makes the interesting assertion that Blacks benefited disproportionately during the Reagan years. His premise is that blacks gained 15 percent of the new jobs under Reagan, though they are only 11 percent of the working-age population.

While these accomplishments and improvements are inspiring, it seems that Reagan has been given more credit than he deserves. Writing in response to Perkins, Douglas Glasgow and Betty Collier-Watson, both of the National Urban League, contend that Perkins exaggerated black progress and failed to connect his claims to the economic practices of the Reagan administration. As far as black employment is concerned, they are not convinced that blacks gained comparatively more new jobs than whites. Blacks did gain 15 percent of new jobs even though they comprise only 11 percent of the working-age population, but Glasgow and Collier-Watson argue from a different vantage point: that the critical comparison is "the black proportion of new labor market entrants with the black share of new jobs gained" (1988, 93). By their calculation, then, blacks "gained less than proportionately from the new jobs created," constituting 16.5 percent of the new entrants but receiving only

15 percent of the new jobs. Furthermore, the black middle class did expand during the 1980s, but this expansion resulted essentially from the "packaging" of different forms of income or from the wages of several members of the same household and not from an increase either in wealth comparable to that of whites or in the proportion of higher-paid black workers (Harrison and Gorham 1992).

Another contradiction between the Reagan administration's rhetoric and its actions centers on one of its tenets—namely the encouragement of economic self-determination. The contradiction can be seen in Reagan's calling for private sector development on Indian reservations, while at the same time withdrawing funds that were being used for development in the public sector on the reservations. Reagan's "definition" of Indian self-determination, in the words of C. Patrick Morris, "flourished in the media but proved fatally undernourished in content" (1994, 74). For example, Joseph Jorgensen points out that "[t]he Federal budget for Indian programs has dropped consistently since 1981" (1986, 10). He goes on to say the following:

> During President Reagan's first seven years he has persuaded Congress to cut the Bureau of Indian Affairs (BIA) and Indian Health Service (IHS of the Public Health Service) budgets annually, and to dismantle programs to provide on-the-job training, education, health care, and social services. (10)

As further indication of the administration's level of concern for American Indians, consider the following: as of October 1986, Reagan had vetoed seven bills related to Indians, which accounted for one-seventh of all bills vetoed by Reagan (Kaplan 1986). Such acts demonstrate Reagan's active commitment to reducing dependence of so-called special interest groups on federal support. But was he equally eager to support the economic self-sufficiency of these same groups?

The answer appears to be that his drive to enhance private sector development for Indians was far less aggressive. For example, on January 14, 1983, President Reagan created a Presidential Commission on Indian Reservation Economies to advise him on economic issues as they affected American Indians. Yet it was not until March of 1984 that he appointed members to the commission. Moreover, Reagan refused to add money to the $10 million revolving credit fund established under President Carter to assist economic development on Indian reservations (Jorgensen 1986). Morris (1994) offers even more compelling data to support his claim that "[t]he devastating effect of the Reagan budget cuts are still resonating in Indian country" (88, note 3). This approach to handling Indian affairs was consistent with Reagan's appeals for economic independence without regard for how disadvantaged citizens are to overcome the material consequences of past and present discrimination to achieve such independence. It seems that Reagan was only willing to provide safety nets for corporations and upper-class citizens, and not for oppressed individuals or groups.

The Reagan administration contested judicial rulings against employment discrimination, remaining true to its pattern of reversing civil rights gains. More specifically, Reynolds and Smith challenged the legality and constitutionality of affirmative action. To this end, the administration used three main arguments against affirmative action, all of which failed to sway the courts. The first argument was that the Constitution does not allow racial discrimination for any reason, good or bad. Next, Reynolds argued that relief in employment discrimination should only go to persons who could prove they were directly discriminated against. Third, Reynolds also maintained that quotas and timetables are not constitutional, and should be replaced by recruiting measures (Sussman 1984). Even though these arguments failed ultimately, the administration's challenges to affirmative action policies bespeak its activist agenda.

During the eight years that Reagan served as President of the United States, his administration had a strong impact on the welfare of racial minorities in the United States, and the impact was largely a detrimental one. For one, the administration caused a regression in the civil rights struggle by its slow and ideologically conservative appointments to important positions of civil rights advocacy. Second, by failing to duly enforce civil rights laws in education, by withholding financial support for higher education programs, and by supporting racially discriminatory institutions, the Reagan administration contributed to the creation of an unjust political environment. Further, through its cutbacks in government job programs and its attempts to reverse affirmative action laws, the administration weakened the morale and the viability of many nonwhite Americans in the employment arena. Although some nonwhites, especially those of middle and upper classes, saw some positive results come about during the Reagan presidency, those results are more incidental than the negative outcomes, which seem to be more direct effects of Reagan policymaking.

The sociopolitical momentum that built up during the Reagan years would carry the country through the rest of the 1980s and early 1990s, maintained politically by George Bush's "status quo" presidency (Duffy and Goodgame 1992). George Bush seemed to most Americans to be more moderate ideologically than Ronald Reagan. For this reason, civil rights advocates expected more support from him for their interests than they received from his predecessor. Yet at least one scholar has argued that Bush surpassed Reagan in his efforts to reduce the federal role in civil rights. Shull (1993) contends that Bush "was more, not less, rhetorical than Reagan and took more assertive actions (including legislative, budget, and administrative decisions)" (4). Bush's record will be discussed in more detail in later chapters, but for now it will suffice to say that "both Presidents Reagan and Bush left a substantial civil rights legacy, but it was not a kinder, gentler one" (Shull 1993, 229).

Cultural Conservatism in the 1980s

Although the anti–civil rights activism of the Reagan administration may have escalated even beyond the expectations of many on the political Right, its actions were not altogether unsupported by Reagan adherents. The actions were merely an extension of the views held by those who had placed

Reagan in office and a partial indication of the country's political shift toward conservatism. An example of these views can be found in a study completed in 1985 for the Michigan House Democratic Campaign Committee. In "Report on Democratic Defection," prepared by the analysis group of the committee, pollster and analysis group president Stanley B. Greenberg wrote the following: "These Democratic defectors believe government has personally intervened to block their opportunities. Appeals to fairness, opportunity, etc. are now defined in racial terms that have been stripped of any progressive content" (quoted in Edsall 1988, 31). Edsall notes that the support in the 1980s behind the GOP in white neighborhoods in Detroit, as well as in Chicago and Philadelphia, "[was] driven in part by the election of black Democratic mayors" (1988, 30–31).

The conservative mood of the times was also apparent in the cultural arena. America beheld the proliferation of conservative cultural expressions and the reactionary responses to more progressive expressions, all of which reflected the discourses and ideologies circulating through American society. These expressions took several forms, among them movies and television programs. This conservatism also reared its head in public and governmental responses to the visual arts throughout the 1980s, as evinced by heightened levels of criticism of and frequent calls for funding cuts to the National Endowment for the Arts (NEA) and the National Endowment for the Humanities (NEH) (Koch 1998). In 1985, for example, three Republican congressmen from Texas (Richard Armey, Thomas DeLay, Steve Bartlett) supported blocking NEA funding to artists whose work could be considered offensive to the average American citizen (Battiata 1985).[9] Applicable here is Young's (1996) comment that "the illusory sense of a superior, coherent identity needs constant reassurance of its fantasized superiority and centrality" (184), and these cultural expressions and responses worked hard to maintain the privileged position of this threatened conservative identity.

Several writers have examined the relationship between culture and the perpetuation of Reaganite ideologies.

In *Selling Culture*, Debora Silverman (1986) identifies an aristocratic movement in 1980s American culture that relied on a celebration of consumerism and a willful distortion of history. She argues that the Reagans were implicated in this movement not only because its supporters were part of the Reagans' circle of friends, but also because the themes of the movement resonate with those of Reagan's presidency. Relatedly, Sidney Blumenthal (1988) has commented that "[n]ostalgia was at the center of the Reaganite aesthetic, which fed upon the cultural debris of the past" (259–260). In the Reagan public aesthetic (of which the first lady was a chief architect), "the beautiful was the familiar" and "also the expensive" (260–261), recalling a pristine, luxurious Golden Age that never really existed in the first place.[10]

Others have focused more specifically on the manifestation of Reaganite ideologies in media culture. Susan Jeffords (1994), for example, places Hollywood films with "hard-bodied heroes" within the context of "the social and political thematics of the Reagan Revolution" (22). According to her analysis, the story lines, characters, and images of several of the most popular films reflect and promote many of the conservative ideologies espoused by President Reagan. She convincingly traces the changes that the Hollywood "hard bodies" underwent throughout the 1980s and early 1990s, and illuminates how these changes correspond to the national models of masculinity and presidential leadership that prevailed during the same period.

Kellner (1995) has also noted the presence of a Reaganite ethos in such Hollywood films as the first two Rambo movies (1982, 1985), *Top Gun* (1986), *Iron Eagle* (1986), *Iron Eagle II* (1988), and *Navy Seals* (1990). Like Jeffords (1994), he conducts an extensive analysis of *First Blood* (1982), perceiving the character of John Rambo to be "a figure of individual entrepreneurialism" (66) and a hero for the Right wing. In a related vein, Kellner (1990) makes explicit television's role in the "creation of a conservative hegemony during the 1980s" (133), and at one point he describes the Reaganite ethos in numerous popular network television programs of the 1980s. He suggests that this can be attributed to the networks' assumption of the popularity of Reagan's conservative mind-set, given how popular Reagan was among American citizens during a large part of his term in office.

As a last example, Herman Gray (1995) comments on the complementarity of Reagan's leadership, the discourse of Reaganism, and television. In short, Gray explores in detail the transformations in the representation of blacks on television during the 1980s and afterwards. The author notes that Ronald Reagan, like Donald Trump and Leona Helmsley, exemplified the conspicuous displays of affluence and consumption that for many people characterize the 1980s. Curiously, race and television played key roles in this conspicuousness: race, in that poor African Americans and other nonwhites served as counterpoints to this affluence, and television, in that it implicitly endorsed the conspicuous consumption by broadcasting it to the country and to the world. Again underscoring the connection between race, politics, and the media, Gray asserts the following: "Race and television were at the very core of the new right's largely successful efforts to establish a rightward shift in the political, cultural, and social discourse [of the 1980s]" (15).

The cultural and political move in America toward the ideological Right did not come to a halt when Reagan left the White House, nor did George Bush's failure to win a second term guarantee that the reign of the right wing was over. President Clinton's win in 1992 did not necessarily indicate the end of widespread support for conservatism. Based on Schneider's (1992) analysis of exit survey data from the 1992 election, the outcome indicated voters' hopefulness that Clinton would help get the economy moving rather than a nationwide "lurch" to the Left. In fact, a majority of polled voters favored a "government that cost less in taxes" but provided fewer services, and that encouraged traditional family values. Support for the Reagan–Bush dead horse of less government, lower taxes, and traditional family values seemed very much "alive" (2542).

In the years since Reagan, several racialized and explicitly political events have taken place; the hearings for Clarence Thomas's appointment by Bush to the Supreme Court constitute one memorable instance. The confirmation process caused many Americans to examine their biases, their allegiances, and their politics. November of 1993 marked the end of congressional hearings on modifications to the racial categories of the U.S. Census. One year later, our country witnessed the Republican Party win a majority of seats in both houses of Congress. Among the party's legions was Governor Pete Wilson of California, whose Proposition 187 called for a ban on education and health care for illegal immigrants. With Newt Gingrich and Bob Dole at the helm, the party pledged a national "revolution" as prescribed in its "contract with America." Shortly before the elections—and not coincidentally—October of 1994 brought with it *The Bell Curve* (Herrnstein and Murray 1994), a neo-eugenicist diatribe intended to provide justification for the race, class, and gender hierarchies that exist in America.

Since 1995 began, many lives have been affected by the wholesale slaughter of women, men, and children in the Oklahoma bombing, committed by members of a white militia group (but initially blamed on foreign terrorists); by the University of California Board of Regents' decision to eliminate affirmative action and the use of race and gender in admissions; and by the surprisingly inflammatory responses to the verdict in the O. J. Simpson criminal trial. And, despite all of these racialized occurrences, Dinesh D'Souza (1995) unapologetically claimed that we have reached "the end of racism." So in the 1980s, at the height of the proliferation of the conservative discourses of individualism, militarism, and laissez-faire economics, and the decline of federal support for public education, Hollywood began to give birth to the films under analysis.

I have devoted a lot of space to discussing the cultural and political impact of Reagan's policies and ideologies. But this study is not primarily about Ronald Reagan; instead, it is centrally concerned with the commonalities and linkages among the policies enacted by the Reagan administration, the American mood since Reagan's first term in office, the messages imbedded in the Hollywood school films under analysis, and the racial positionalities of the characters in the films. Traversing the various neat disciplinary boundaries that would normally divide such subject matter into several parts is a useful approach in order to better comprehend the ideological work involved in these Hollywood texts. And a consideration of how the Reagan legacy is implicated in these texts is indispensable. The studies of Kellner (1990, 1995), Gray (1995), Jeffords (1994)—among others—powerfully testify to this fact; consequently, they serve as models for the attempt in this study to uncover how the political, cultural, and racial are interwoven in the sample of school films.

If, as Michael Rogin (1987) argues, Ronald Reagan's presidency reflected the Hollywood movie roles that he played, then it seems reasonable to look to the movies for resonances with his leadership. I purport that the politics of the films in question—starting with *Class of 1984* and ending with *Music of the Heart*—do resonate with the ideologies and discourses that have circulated since Reagan. Some of the films are more closely or directly linked to policy examples than others, but all of them are tied to the Reagan era in some significant way. I will therefore attempt to make some connections between the presence of these Reaganite ideologies in society and the representations in the films, specifically as they concern race.

In the following three chapters, I try to make sense of the films by grouping them categorically. Of course, the categories overlap to some degree, and all of the films exhibit characteristics of each category, but I divide them according to what I see as their predominant messages and themes. In addition, the manifestations of the Reagan legacy in the eras of presidents Bush and Clinton will be discussed in these chapters when it is pertinent to the analysis at hand. With that to look forward to, let us now delve into the body of school films.

3

Controlling the Savage Enemy: Punishing the Other and Criminalizing Urban Space

For many Americans dissatisfied with the leadership of President Jimmy Carter, Ronald Reagan must have appeared as a "macho" (Orman 1987), swashbuckling hero right out of a Hollywood movie, ready to save America from its fiercest enemies. These perceived enemies as constructed by the Reagan campaign were simultaneously foreign (Communists, terrorists, and Latin American dictators) and domestic (crack addicts, welfare queens, and gang members) or in the case of "illegal aliens," somewhere in between. While these threatening enemies were of various nationalities, they were almost always dark complexioned. Reagan's presidential mission, as he and his supporters saw it, was to rescue the country for the average, taxpaying, American citizen; in decoded words, for middle-class whites.[1]

Several actions pursued under Reagan's leadership fulfilled his promise to rid America of the specific dangerous "Others" mentioned above. The threats from outside were met with assertive responses: the acceleration of the nuclear and space-based arms races; the denouncement of Iran and Nicaragua as sponsors of international terrorism; the "wars" waged on the battlefields of El Salvador and Guatemala; and the use of U.S. refugee policy as the handmaiden of foreign policy. The Reagan administration's responses to domestic threats were equally aggressive, exemplified by the so-called war on drugs, the severe cutbacks in government programs that benefited the disadvantaged, and the construction of more prisons and increase in the United States incarceration rate (LaFraniere 1991), all of which disproportionately affected poorer, nonwhite citizens.[2]

Coupled with this wholesale attack on the nonwhite Other was the circulation of the view of the city space as a place to be feared, avoided, or controlled. "Dramatic new patterns of segregation have emerged," argue Omi and Winant (1993), "as whites have moved to the suburbs surrounding the declining cities, many of which have majority black and Latino populations" (102; see also Edsall and Edsall 1991). These patterns of white flight are undoubtedly a continuation of the expansion of suburban communities since World War II (Jackson 1985). Many whites have by now fled to gated suburban communities perceived as havens from an unsafe urban environment (Christopherson 1994). Edsall and Edsall (1991) point to the consequence of such a negative view of city life in their claim that the nation is moving steadily toward a national politics that will be dominated by the suburban—largely white—vote. They suggest that the "white noose" effect (230) created by the expansion of Republican suburbs surrounding particular inner cities chokes the life out of these urban areas and leaves them virtually dead.

Popular media imagery reinforced this dual fear of nonwhites—especially blacks and Latinos—and of urbanness. In the realm of television, images of undocumented immigrants (particularly from Haiti and Latin America), and of blacks competing for decreasing resources and job opportunities (especially during the recessionary period of 1982) did nothing to allay the fears of middle- and working-class whites (Gray 1995). Equally unenlightening was the trend of local media outlets toward featuring prominently in their news programs violent crimes committed against innocent (often white) victims at the hands of young, black, male youths. "Together with these racialized images of the 'immigration problem,' " Gray (1995) writes, "news footage of drug and violence-ravaged urban communities made it seem as if (white) America were being attacked from within and without" (20; see also Reeves and Campbell 1994).

Perhaps the most memorable image encapsulating the right-wing desire to both punish nonwhites and demonize the city space is the Willie Horton advertisement employed by George Bush's presidential campaigners in 1988. Horton, of course, was a convicted murderer who raped a white woman while on furlough granted by Massachusetts Governor Michael Dukakis, who was Bush's opponent. For anyone searching for justification of and an explanation for white flight, the Willie Horton experience provided it.[3]

The literal and symbolic association during the Reagan era of nonwhites with the perceived criminality and squalor of the city space is not exactly without precedent in American cultural texts. During the period following World War II, for example, advertisements often featured suburban schools with desirable playgrounds and facilities, whereas their urban equivalents were photographed with padlocked doors and barred windows (Jackson 1985). More specifically in terms of images in school movies, *Blackboard Jungle* (1955) is the most obvious Hollywood prototype dealing with "urban school problems." The film certainly plays on the fear of city violence (before Los Angeles replaced New York as the symbolic locus of urban decay[4]) and inserts a racial element into this fear. However, it seems less concerned with punishment and focuses more on the appropriation of the racial Other to solve its central dilemma(s). In the end, *Blackboard Jungle* chooses to depict the nobility of the feared black student's choice to support authority instead of resolving the issue through force, punition, or coercion of his black body.

The main reason the attempt to connect cities, violence, and nonwhites becomes notable during the Reagan era is that these elements were (and continue to be) presented as inseparable. After the vocal demands of the civil rights era, there was a quiet interim of "harmless" school films about teenage concerns (*Grease* (1978)), sex and good-natured rowdiness (*Animal House* (1978)), the vicissitudes of academic life (*The Paper Chase* (1973)), and the well-intentioned crossing of racial boundaries (*Conrack* (1974)). However, with the election of Ronald Reagan and the beginning of the conservative revolution, the 1980s gave rise to a number of movies that portrayed a much more "harmful" threat to the social order: the rampant, "colored" criminality of schools in the central city. Under Reagan, so soon after the civil rights movement, it became culturally acceptable for whites to feel less guilty about perceiving nonwhites to be a threat to social and economic stability. By either producing films that dealt with their fears or identifying (as audience members) with the messages these films conveyed, many whites could safely express their racial frustrations and concerns about their own well being without appearing racist (Gray 1995). Common sense tells us, after all, that films are just entertainment.

The Urban School Problem on Film

There were a few 1980s Hollywood school films that took place in inner-city educational institutions that did not explicitly connect nonwhites to the criminality of these institutions. These films actually placed white characters at the center of their narratives, allowing for the interpretation that whites could be criminal, too. One such movie is *Class of 1984*, which has been classified as both an exploitation film and a cult movie. However, it can be considered a school film for the purposes of this study because it focuses to a great degree on the educational nature of its setting more than a horror flick like *Sorority House Massacre*. Another such film to be discussed here is *Teachers*. For all of the differences between these movies, I will eventually comment on the similar linkages they make between race and urban educational institutions.

Although *Class of 1984* draws upon earlier horror, drama, action, and school films for inspiration, it is readily apparent that the film relies on the exploitation of violence for much of its energy.[5] The issue of violence is immediately emphasized in the opening intertitle of the film, printed on the screen in white letters on a black background:

> Last year there were 280,000 incidents of violence by students against their teachers and classmates in our high schools. Unfortunately, this film is partially based on true events. Fortunately, very few schools are like Lincoln High . . . yet.

It is not apparent whether the phrase "partially based on true events" is accurate or whether it is merely being used to set the mood for the film or both, since this is the only indication the filmmakers offer that their narrative may have some basis in real life. In any event, what is apparent is that there is an intentional connection between violence and the urban, a connection that is meant to incite fear among the viewers.

After this introductory text, the movie opens with shots of the campus of Abraham Lincoln High School as well as the unnamed city in which it is located. The audience is treated to glimpses of the vice of this locale through shots of the vandalization of a building and of a sign in the school's faculty parking lot, of a group of students smoking marijuana, and of a breakout of student aggression on the schoolyard. A veteran biology teacher, Terry Corrigan, pulls into the campus parking lot, exits his vehicle, and checks his briefcase, revealing a gun. These images convey rather quickly that crime, at least in this city, is everywhere.

Lincoln's newest teacher, Andy Norris, observes Corrigan's gun as he passes him in the parking lot. Corrigan catches up to Norris and initiates the following conversation:

CORRIGAN:	I'll bet I know who you are. You're Goldstein's replacement. I'm Terry Corrigan, biology.
NORRIS:	Andy Norris, music. What's the gun for?
CORRIGAN:	Where have you been teaching lately?
NORRIS:	Lately, nowhere.
CORRIGAN:	It figures.

The two teachers walk into the school together and Norris immediately expresses disbelief at the sight of metal detectors at the student entrance. The first student to set off the detector and be frisked is a member of a black gang, identifiable by his head accessory. Norris asks Corrigan, "Is that really necessary?" "It is if you want to survive," Corrigan responds matter-of-factly. Afterwards, Norris visits the principal to obtain his teaching assignment, and finds him preoccupied with his video monitors. "Surveillance is the name of the game, Mr. Norris," the principal tells him. The film then underscores the connection it has been striving to make between the criminal and the urban in the principal's next statement: "You're not in Nebraska anymore, Mr. Norris."[6]

Not long after his arrival at the school, Norris encounters some disciplinary problems in his music class from a few students who have been established early on as the school toughs. After ousting them from his classroom, Norris later finds the leader of the pack—a white youth named Peter Stegman—and his henchmen in a bathroom just after the completion of a drug deal. Norris reports the incident to the principal and Stegman dissembles and claims to be the victim of Norris's in-class discrimination. To Norris's surprise and chagrin, the principal sides with the students due to a lack of evidence. By now, the specter of criminal behavior at Lincoln has been made patently visible.

The antagonism between the faculty and these students—particularly between Norris and Stegman—continues to swell throughout the movie, mediated by the principal and, later, a local police detective who are both depicted as protecting the students' legal status as juveniles. The battle becomes progressively more dangerous and more personal, and climaxes with the rape and kidnapping of Norris's wife by Stegman's gang on the night of the school orchestra's big concert. One of the orchestra's most talented members has to assume the responsibility of conductor because Norris is occupied with hunting down and killing each of the gang members and rescuing his wife from Stegman. All of this retribution occurs unbeknownst to the principal, students, or parents at the concert. After all, Norris was unable to garner the support of the authority figures in his fight against these students prior to this point, so he ultimately has to expel this criminal element from the school and the city all by himself.

It is necessary to state at this point that Stegman and all of the members of his gang are white, and that together they pose more of a violent threat than any students depicted in most school films before or since. On the surface, this fact would seem to cause *Class of 1984* to stand out as a departure from the "nonwhite students + urban institution = violent threat" formula. Indeed, the movie is rare among its counterparts from the Reagan era and beyond for this distinction.

With *Class of 1984*, however, as with the other Hollywood school movies in this study, the messages of the text are hardly guaranteed by the race of its principal characters. Hence, there are a few important points to be made about the connection between race and criminality in this film. First, although all of the students in the Stegman gang are white, they are not just "average" white students, but constitute a special, isolated subset of whites in the school. The majority of the other students at Lincoln High (most of whom are white) appear to be unconnected to the criminal activities committed by Stegman and his pals. In fact, the white young men and women in Norris's music class are either resentful of Stegman's presence and/or intimidated by his tactics. Norris and the principal of the school point to the gang's deviance near the end of film when Norris says, "There're a lot of good kids here." The principal responds, "I suppose so. It's just that the bad ones take so much of your time."

A second yet related point is that the film's narrative goes out of its way to demonize the white gang in order to place them outside of the mainstream of the school's population. They are, in other words, portrayed as a sort of radical fringe group of whites at odds with the rest of the students. One way in which this is accomplished is through the identities the characters are assigned. Judging by the clothes they wear, the music they listen to, and the establishments they patronize, it is evident that the film places the gang members securely within the punk rock culture of the eighties. The gang members' use of language and symbols also contributes to the unflattering identities they bear in the film: they use the word "suburbanite" as an insult; on one occasion they refer to black people as "niggers"; upon being ousted from class by Norris, they make a hand sign reminiscent of a "Heil Hitler" salute; and two of them wear t-shirts with a swastika on the front. On top of that, they sell drugs inside and outside the school, and they run a prostitution ring. In short, the film effectively marginalizes the Stegman bunch and makes a strong case that these students are not to be confused with the typical Lincoln High student. In highlighting the gang's deviance from the mainstream, the film suggests that the problem with these students is *individual* and not *racial*, and they are drawn to each other because of their shared individual problems. This is unlike what we shall see in films that deal with delinquent nonwhites, in which just the opposite is usually suggested. The message here is clear: these kids are not just your garden-variety teenage delinquents; they are a very small, very radical faction of students who endanger the educations and lives of the real students.

The third point to be made about the racialization of violence in *Class of 1984* is that, despite the explicit connection between the presence of the white gang and the existence of violence at Lincoln, the white gang members are not the sole perpetuators of this violence.[7] In three instances in the film, fighting breaks out involving the white gang members, and in all three instances the opponents are predominantly black. In the first case, Stegman and his gang punish a black student for selling drugs at school—on their turf. As a result, the drug dealer's black allies confront Stegman's gang after school and fight brutally. After Stegman's gang gets the better of the black group, the police come and arrest the injured blacks while the white youths escape. In another instance, an initiate into Stegman's gang stabs an innocent white student in the school cafeteria while the other gangsters start a diversionary fight with some random black kids. Thus, even though Stegman's gang of white students represents the principal threat of violence in the film, black criminality is definitely not excluded from the narrative.

Last, the film makes an attempt to redeem the goodness of the white students (and of whiteness itself). For all of Stegman's morally reprehensible and legally prohibited behavior, and though he calls Stegman "psychotic," Norris remains convinced that he is "actually a brilliant kid." As a matter of fact, the narrative of *Class of 1984* seems to point accusatory fingers at the principal who is unwilling to support his faculty over his students, at the police detective who is reluctant to prosecute them, and at Stegman's mother who is overly protective (and dangerously enabling). It is therefore the failure of the principal, the police, and the parent to punish that has produced such a criminal band of white youth. It is not until

Norris takes matters into his own hands that the gang members pay for their crimes. Interestingly, the general release of *Class of 1984* in August of 1982 coincides with a period of national television and newsmagazine coverage that Gilens (1999) demonstrates as featuring more "sympathetic coverage of white poverty" (125). His analysis of media images of poverty shows that in 1982 and 1983, there were proportionally more images of whites among the more sympathetic portrayals of the "deserving poor" than from 1972 to 1973, for example, when the portrayals were unsympathetic and featured very high proportions of blacks (127–132). Connecting race to poverty is not the same as connecting it to delinquency, of course, but the parallel displays of sympathy toward and redemption of whites are noteworthy.

In spite of the seriousness and enormity of the crimes committed by Stegman and his gang, the movie avoids making their racial similarity an explanatory factor for their behavior. The white students in the gang are displaced from the mainstream and depicted as troubled individuals, not racialized subjects. At the same time, the film tries to rescue their racial identities by telling a story of "good kids gone bad." Their behavior is contextualized within a network of deleterious urban influences, mental instabilities, and ineffective authority figures. Moreover, a hint of black criminality is thrown in for good measure. Later we shall see how this recipe for an urban school film with whites as the main ingredient differs significantly from the films that feature racial Others.

Teachers, released in 1984 to large box-office crowds (Sanello 1984), is another film that explores the connection between crime and urban schools. From the beginning of the movie, the audience is exposed to a chaotic public school setting in which unruliness and violence are rampant: the school's hallways are enclosed in cages that have to be unlocked before school begins and remain locked at all times; a black male stabbing victim sits in the school's office without medical attention; a teacher is bitten on the hand by one of his students; an escaped mental hospital outpatient is undetected as he fills in as a history teacher; and, most ironically, the school psychologist has an emotional breakdown that is sparked by a teacher's hoarding of the mimeograph.

The plot of the movie centers on Alex Jurel, a morally spent white male teacher at John F. Kennedy High, and his struggle to regain some sense of vocational purpose in a school being sued for graduating an illiterate student. While the white female superintendent barks orders and the black male school attorney lies, stalls, and conceals in order to save face, Jurel tackles the personal problems of his students and the ugly politics of academic bureaucracy. At one point, he responds to his situation by saying, "I try to stay out of political things. I have a hard enough time just teaching." Ironically, Jurel was once a radical, caring, politically involved teacher, but his disillusionment with the state of schooling overtook his motivation and radical leanings. He eventually reclaims some of that initial motivation, however, as he accepts responsibility for the welfare of his students and becomes personally involved in their lives.

The film highlights only a few of the students at predominantly white Kennedy. A white male student is depicted as edgy and possibly schizophrenic. One of Jurel's students, a black male, is clownish and garrulous. Later in the film, one of his white female students becomes pregnant by a physical education teacher at the school. Jurel becomes embroiled in controversy when his noble yet inappropriate action of taking the student to have an abortion is discovered. Jurel's most challenging student, however, is Eddie Pilikian, a white youth who is on his way to becoming one of the "illiterates" for which Kennedy High is notorious. His deviant behavior includes stealing an exam, taking a car from the driver's ed fleet (which he repairs and returns), and setting off fire alarms. Jurel finds him to be resistant to any helpful overtures, but he finally convinces him to retake remedial reading (upsetting his self-interested and obstructionist parents) and wins his confidence.

For all of their shortcomings, though, Jurel's students are not violent or excessively criminal. The most delinquent of them is Pilikian, and as one critic points out, his "only problem is that he is 'misunderstood,' and his goodness becomes apparent too soon after he is 'understood' by an authority figure who listens to and trusts him" (Edelman 1985, 11). If all of the students are like Jurel's, one wonders why the cages and security guards in the school are necessary. In sum, there is an obvious presence of crime within the school, but little evidence of any perpetrators. The only persons we see explicitly connected to a violent act are the black male student who is stabbed at the beginning; and Pilikian's friend, who bites a teacher, gets pummeled by a group of leather-clad whites, and is later accidentally and fatally shot by the police during a locker search. If the students are a bit wayward, it seems that it is the result of deplorable influences incarnated in their parents and other irresponsible authority figures.

I want to point out a few similarities between the two films discussed above that are germane to themes of this chapter. Despite their obvious differences of subject matter, commercial success, and treatment, *Teachers* and *Class of 1984* feature students, mostly white ones, who pose a significant threat to the social order. Disorder, violence, and crime are enormous problems in the schools in both films. Due to the failure of the authority figures to handle the criminal element by meting out the appropriate punishment, it is up to some lone individual to restore some sense of order in the school. This lone individual in both films is a caring, white, male teacher, which is interesting given the fact that about 70 percent of public school teachers in the real world during the eighties were women (U.S. Department of Education 1996).[8]

Another element the two movies share is that the high school in each movie is named after a former United States president. This, of course, is not uncommon in films; however, the importance of this element lies in the specific names used and their connection to what the films are about. Perhaps the filmmakers were counting on the potential information communicated by the legacies these names might invoke: one (Lincoln) being that of a honest "emancipator"; the other (Kennedy) recalling the liberalism of the sixties; and the fact that both fell victim to acts of violence. The suggestion to the spectator might be that the poor parenting, drug abuse, violence, and obsession with sex that resulted from liberal leadership have defamed the names of these great presidents and are responsible for the decline of today's schools. The effects of this liberalism could be counteracted by family values, school prayer, and militaristic discipline—in other words, the platform of then-President Reagan, who had himself already been the victim of an assassination attempt by the time these films were released.

One last commonality is that those students associated with violence and other crime in the two films are treated differently in the narratives depending on their race. When white students are assigned blame for the crime that exists in these schools, they are portrayed as troubled individuals whose behavior is a result of their social conditions or mental health and not their racial backgrounds. When the criminals are African American, no attempt is made to contextualize, understand, or excuse their actions. In short, the delinquents are either emotionally disturbed whites or they are black. These racialized depictions correspond with what Reeves and Campbell (1994) report as perhaps the central finding of their study of the television news coverage of the "anti-cocaine crusade": "the disparity in news treatment of 'white offenders' and 'black delinquents' " (42), indicating that the media representations place emphasis on the *actions* of whites as opposed to the *pathology* of blacks. The importance and relevance of this last aspect will become more apparent as I examine later school films whose cast of characters are more racially diverse. In these "multiracial" films, urban violence is disconnected from socioeconomic conditions and directly linked to the nonwhite individuals themselves.[9]

Urban School Violence in Multiracial Films

The theme of punishment becomes much more salient in *The Principal*, a school film in which the cast is more racially diverse and whites are in the minority. One might speculate that the makers of *The Principal* and successive Hollywood school films moved toward more racially diverse casts in order to attract the large number of African American moviegoers—commonly estimated at one-fourth to one-third of the paying domestic audience despite comprising only 12.3 percent of the U.S. population (Rhines 1996; U.S. Census 2000)—while retaining a significant white audience as well. Thus, many of them appear to have relied on an old, established narrative—that of *Blackboard Jungle* and of the more recent *Conrack*. In other words, the filmmakers offered a conservative solution to their central dilemmas by rebuilding "older practices and paradigms" (Guerrero 1993, 122). The problem here is that American society had undergone sea changes since those earlier films debuted on the silver screen. This recalls Snead's (1994) perceptive claim that "racism in the cinema might be described as the tendency to recycle certain ethnic codes, already familiar to a series of privileged viewers, in order to reinforce their familiarity, despite the changes that may have gone on in the real world" (142). The films certainly reflected the contemporary environment, as evidenced by their settings. Yet, while the complexion of the problem had changed somewhat, the complexion of the solution remained the same.

The Principal, as one reviewer has suggested, looks like a cross between *Rambo* and *Blackboard Jungle* (Lumenick 1987). Given the public's extreme familiarity with these two films, this description may for many evoke an image of an imperiled American institution housing a valiant teacher and uncontrollable students and rife with violence, punishment, hypermasculinity and, in the end, triumph over evil. All of these elements are in fact present in the movie. In addition, it represents a departure from earlier school films in that blacks, generally speaking, are the principal threat in the school. Importantly, this was one year after television news coverage of crack cocaine had reached a turning point—a point at which the coverage became racially coded (Reeves and Campbell 1994).

The premise of *The Principal* is that teacher Rick Latimer loses his job at a suburban high school in southern California as a result of his behavioral problems. In a drunken rage Latimer attacks the Porsche of his ex-wife's divorce attorney—with a bat, no less. Somehow the personal and the professional become entangled when this brash act becomes the ultimate factor in the administrative decision to banish him to Brandel High. Apparently, the school's disagreeable reputation registers immediately with Latimer as he reluctantly accepts the position of principal of Brandel.

As Latimer travels on his motorcycle from his neighborhood of residence to the location of the school, the film makes the audience aware of the stark contrast between the pleasant suburban atmosphere of his home and the urban blight in which he must now work. Although it is evident that he is not pleased with the school's surroundings (he derisively calls it "Disneyland"), it is not because he is some sheltered, timid suburbanite afraid of working with the school system's rejects. Considering the manner in which he harasses his ex-wife's attorney and given his chosen mode of transportation, the film invites the audience to view Latimer as a tough guy who is ready to face the challenges ahead of him at Brandel. This tough guy image is soon confirmed when he chases after a carload of African Americans pursuing three white youths on foot, by which he prevents some act of crime from occurring. He arrives just in time to break up a potentially serious altercation and carries the two combatants (one black and one white) into the school, where he then learns that they are both Brandel students who are affiliated with rival gangs. Thus, in the first 15 minutes of the film the putative connection among nonwhite students, urban schools, and violence is again reinforced, and this connection sets the stage for viewing the rest of the film.

A survey of the entire film reveals that the components necessary to make the above connection are present. First, the audience is bombarded with the familiar litany of disparaging comments about the students: "truants," "thugs," "animals," "garbage." These descriptors are not meant to refer to all of the students, of course, or perhaps not even most of them. But the pall cast over the school by its most dangerous students overshadows the efforts of the obedient, conscientious ones. Second, spectators are presented with an intimidating yet caring hero—a white man—who is a fearless force to be reckoned with, but not too busy to help those students who need him most. Third, the film also includes a trusty sidekick, the black male head of security, a former football star at Brandel who seems to know enough about the school to be able to run it effectively but is curiously content to assist the principal in doing his job. The principal obviously relies heavily on his loyalty and support and *brawn* throughout the movie, but in the end the principal rids the school of its most putrid "garbage" all alone. Last, there is the presence of the threatening Other—in this case black student Victor Duncan and his gang of thugs who use the school to conduct their illegal business.

It is interesting to note the race of the film's antagonists given that the student body at Brandel appears to comprise a relatively equal mix of blacks, Anglos, and Hispanic Americans. One of the primary delinquents in the school is the white youth whose life Latimer saves at the beginning of the film. His membership in the white gang is problematic enough, but his behavior becomes more serious when his darker-skinned Latina history teacher embarrasses him in class and he exacts revenge by assaulting and attempting to rape her. She is rescued by Latimer, and the white youth is arrested.

Yet, the one student who poses the most significant threat to Latimer and the rest of the school is Victor Duncan. He is a prosperous drug dealer who sees himself as the de facto head of Brandel and expects the principal to work in peaceful coexistence with him and his black "assistants." What is more, Duncan plays a part in almost every criminal act that occurs on the grounds of Brandel. For example, Duncan employs a black student-mother to sell drugs and perform other jobs for him, like setting up Latimer to be attacked by Duncan and his gang. He is also responsible for seriously injuring another black student who quits his gang. Last, Duncan assisted in the assault of the Latina teacher because he was interested in recruiting the white perpetrator to work *against* the white gang and eliminate them. These examples indicate the extent of Duncan's influence.

At the end of the film, Latimer is faced with the potentiality of having to fight for his life against both Duncan's black gang and the solo white gang member (who has been released from jail earlier that day). But with the assistance of his black sidekick and a loyal student, the principal is able to handle the danger with which he is confronted. After all the other antagonists have been eliminated from the picture, the film underscores who the real enemy is with a showdown between Latimer and Duncan. The narrative ends with Latimer literally throwing the gang leader through the school's front doors, proclaiming "No more!" and reclaiming Brandel High for the "real" students.

Generally speaking, *The Principal* portrays its characters along a continuum of delinquency, which might look something like the following:

most delinquent *least delinquent*
<---------black students--------Latino students--------white students--------->

The majority of the student troublemakers in the film, including the student-mother, Treena, are African American. Of this majority, only a few appear to be conscientious, bright, or otherwise "redeemable." The school's Hispanic population is composed of both productive (Arturo, Raymi, the auto mechanics) and destructive (miscellaneous gang members) elements. As for the white students, there is mention of a white gang, but the audience learns about only one of its members, and he is the only identifiable white delinquent. And, again, Duncan is the most powerful criminal presence at Brandel. Other students are guilty of rebellion and delinquency in the film, but their actions are secondary to Duncan's criminal behavior.

The Principal appears to have started a trend in school films toward featuring the criminal element as nonwhite, a shift away from offering explanations for wayward behavior and a mere acceptance of it as an inevitability, and the emergence of the rhetoric of individual responsibility. The next school film to be considered, *Lean on Me*, picks up where *The Principal* leaves off and covers new ground through the spin its filmmakers put on the "urban school problem" narrative. I examine this film in more detail than the others because of the remarkably close resonances between Reaganism and the film's messages. With its bat-toting principal and black antagonists, *The Principal* paved the way for the Hollywood release of *Lean on Me*.[10]

The Case of *Lean on Me*

In January of 1989, the United States of America was emerging from eight years of direction under Ronald Reagan, a socially conservative time period not at all kind to many citizens, especially poor African Americans. As the presidential baton was being passed to George Bush, we were promised a "kinder, gentler" nation. Two months later, in the midst of this changing of the conservative guard, the film *Lean on Me* opened as the top box-office draw in Hollywood. The film recounts the story of Joe Clark, the real-life principal of Eastside High School in Paterson, New Jersey, who dramatically transfigured the chaotic, depressed school into a tidy institution of order and discipline. The epic tale traces Clark's embattled journey to success at Eastside as his students strive to pass the state's minimum basic skills test. As the principal employs controversial disciplinary measures to bring the school under control, he is both challenged by officials with selfish interests and encouraged by supportive parents and students. Just when the situation looks its worst, the film rescues Clark from the cabal of local leaders and presents him as a hero.

In its celebration of a heroic male savior, *Lean on Me* resembles other Hollywood school films—in this sense, it offers little new thematic ground to explore. What causes the film to stand out from the cluster of other films discussed in this chapter, though, is that it heroicizes an African American man during a decade when African Americans were being dehumanized in mainstream popular cultural texts. The film's release is even more anomalous given the amazing dearth of Hollywood films primarily about blacks or featuring blacks in starring roles (Guerrero 1993). The existence of these ostensible paradoxes leads to the central question in the examination of this film: Why did the filmmakers of *Lean on Me* decide to tell the "positive" story of a black man, and why was Joe Clark chosen to be that man? In short, how did *Lean on Me* become a Hollywood school film?

The following examination attempts to uncover the agenda of its filmmakers—specifically, producer Norman Twain, director John Avildsen, and screenwriter Michael Schiffer—and the manner in which they created a text whose underlying messages epitomize the conservative sociopolitical mood of the Reagan era. Through their use of fabrication, revision of history, reliance on racial, gender, and class stereotypes, and incorporation of Joe Clark's rhetoric, the filmmakers produced a work that embodies the conservative discourses circulating in America in the late 1980s. I believe the film's interest as an object of analysis lies in the combination of its "feel-good politics" with the promotion of various other elements: Reagan-era traditionalism, conservative educational prescriptions, individualism, unfettered free-market competition, and classist stereotypes.

Lean on Me is the filmic account of how Joe Clark, at one time a teacher at a New Jersey high school, returns to the school as its principal and transforms it from a drug-infested, violence-ridden place into a safer, quieter institution of learning where strict discipline and clean hallways are the order of the day. The textual introduction to the film offers the following synopsis:

> The following is based upon a true story. Once considered among the finest high schools in America, Eastside High of Paterson, New Jersey, declined over the years until an official report called it a terrible "cauldron of violence." The battle of one man, Joe Clark, to save Eastside High School and restore its former pride is the subject of our story. It began about twenty years ago.

At the center of the film's story is the struggle by Clark, his teachers, and his students to perform at a certain level on a statewide standardized exam or else lose local control and fall under the aegis of the state. In order to achieve this, the school board hires Clark to assume the formidable task of raising the students' test scores within the course of one school year. The film's dilemma arises when the mayor (egged on by a surly African American mother and a bigoted white fire chief) challenges Clark's methods for attaining order in the school. In the end, the indefatigable Clark wins the broad support of students and parents alike and the required percentage of students pass the standardized test.

The film was directed by John Avildsen, probably best known for his work on *Rocky* (1976) and *The Karate Kid* (1984). Although the subject matter of *Lean on Me* is markedly different and inspired by real-life events, the director relies heavily on the same formula as in the earlier "fighting" films. Several of the film's reviewers highlight its resemblance to Avildsen's other works, including one who believes that the director "is a master at manipulating audience response through his depictions of men (or boys) struggling against great odds" (Marcazzo 1990, 210). In order to make his formula work this time around, Avildsen employs Morgan Freeman as the "pedagogical Rocky" in order to capture the essence of Clark. As Clark, Freeman is an energetic disciplinarian, and he is convincing in his delivery of Clark's notorious polysyllabic vocabulary. But how much is his character based on the real Joe Clark?

Will the Real Joe Clark Please Stand?

Lean on Me begins in an Eastside classroom in 1967, with Clark conducting a history class comprised almost exclusively of white students. The dashiki-clad Clark is shown engaging his students in an academic game, surrounded by a room full of civil-rights-era memorabilia. In the middle of this educational exercise, Clark's friend and colleague Frank Napier (who eventually becomes Paterson's superintendent of schools) interrupts him and warns him that the other faculty members are in the process of voting to oust him from Eastside, presumably because of his radicalism. Upon being transferred from the school, he loudly proclaims, "This place deserves exactly what it gets!" Not surprisingly, his curse works twenty years later as we witness the screen dissolve to the new (but certainly not improved) Eastside, with its graffiti-filled walls, violently abusive students, fearful teachers, and rampant drug-dealing. As if the visual depiction of the transformation were not enough, the scene is milked of all its melodramatic potential by the addition of Guns 'N Roses's "Welcome to the Jungle" playing in the background. This is the Eastside that Clark is called in to lead.

However epic Clark's life may be, this dramatic account of it is apparently far from accurate. According to Denise Gellene (1989), who knew Clark when she worked as a reporter for the *Record* of Hackensack, New Jersey, the educator did not begin teaching at Eastside until 1972. Clark taught remedial reading and his students were mostly black. He was in fact sent to an all-white elementary school, after which he eventually became principal of School No. 6, a substandard black elementary school near Eastside. He reportedly improved the school's academic record drastically during his six years there. Thus, Clark had countered the view of him as an "ideological troublemaker" and was recruited for Eastside (Chapelle 1989; Gellene 1989).

What the film is more accurate at depicting, though, is the magnitude of the situation Clark faced once he was hired as Eastside's principal. When he arrived in 1982, the high school was comprised of 3,200 students, the majority of whom were poor and working-class African Americans and Hispanic Americans. In the 1979 report that labeled Eastside "a cauldron of terror and violence," the county prosecutor discusses the beatings, vandalism, drug use, and theft that existed there in crisis proportions. It was to this unconducive atmosphere for learning that Joe Clark was called to restore order, to raise test scores, and to lift spirits.

Perhaps it is at this point that the filmic account of Clark's life begins to diverge from his real-life experiences. This is not to say that the narrative of *Lean on Me* is totally at odds with reality, for the film largely resonates with the style, behavior, and approach that Clark exhibited while at Eastside. And, after all, it is neither unusual nor criminal for Hollywood films or any artistic form to take liberties in its representation of "true" stories or real life. Yet there is more to the portrayal of Clark's life in the film than artistic expression. I want to argue instead that the filmmakers, buttressed by the conservative mood of the Reagan era, use *Lean on Me* to promote their right-wing agenda. The film is, in other words, a means of disseminating their conservative views—as well as those of Clark—on race and education in the sociopolitical climate of the United States in the late 1980s.

The Reagan, Bennett, and Clark Connection

Irrespective of the accuracy of the film's portrayal of Clark, there is one factor on which the film does not focus that is essential to understanding how his story emerged from Hollywood within the political and cultural environment that was contemporaneous with its release. That factor is Clark's ideological connection to the Reagan administration. The following statement by Jay Carr (1989) alludes to this connection: "The script to *Lean on Me* plays like something written by the Reagan administration. It supplies a rationale for white-controlled governments to ignore the educational needs of largely black schools districts that need funding most" (43).

The fact that the narrative of *Lean on Me* resembles a Bennett prescription for educational improvement is no accident.[11] Given that Bennett considered Clark to be a friend, one might expect them to be ideologically similar. Both Bennett and Reagan have been publicly adulatory of Clark's pedagogical accomplishments on more than one occasion. For instance, the former principal received a commendation from the White House for his work at Eastside. He was even offered a job in the wake of a 1989 controversy at his school as an advisor to an advisor to the president. And in several ways Clark's philosophies dovetail with those of Reagan and Bennett.

Take Clark's views on the role of discipline in education, for example. As he sees it, "Discipline is the ultimate tenet of education. Discipline establishes the format, the environment for academic achievement to occur" (Chapelle 1989, 68). Clark's focus on discipline above all else is hauntingly similar to the Reagan rhetoric that those who labor in inner-city schools have to "get tough" in order to succeed. This disciplinarian outlook is also urged on by views like Bennett's, who told the press that in leading the nation's schools, "[S]ometimes you need Mr. Chips, sometimes you need Dirty Harry" (Friedrich and Bowen 1988, 53).

Clark's views on the status and utility of public education also match the ideologies of the Reagan administration. At a lecture in Tacoma, Washington, Clark characterized the urban education system as "rotten to the core," with no room for improvement. He said that urban education needs to be enhanced by an open enrollment or a type of voucher system so that parents may send their children to any public or private school they choose. And he also complained that schools "have become antithetical to the American process of free enterprise," indicating a need for more academic, administrative, and athletic competition between schools (UPI 1989). It is anyone's guess whether the principal actually shared the same views as the Reagan administration or was merely playing politics by regurgitating White House policy of the time. The point is that Clark's testimonial fully supports both the original goals of ED under Reagan and Secretary Bell and the president's commitment to free-market competition.

Two other parallels between the agendas of Clark and the Reagan administration are their militaristic approach to leadership and their attacks on drug use. In the first case, the influence of militarism is evidenced in ex-Army Reserve Sergeant Clark's tenet of "discipline above all," as well as in Reagan's obsession with policing and protecting borders. For the former, leadership involves speaking loudly and carrying a big stick; for the latter, war was an overused metaphor for everything from the prevention of illegal drug use to competition in space exploration. As for the second issue, fighting drug abuse has been of great concern to Clark and to William Bennett. Clark, as we know, successfully eliminated the rampant sale and use of drugs at Eastside High. Bennett was rewarded with the position of Drug Czar under George Bush for his preoccupation with illegal drug use by young people. The two men's commitment to reducing drug use and abuse in America is highly commendable. The complaint here is that, while in their positions as educational leaders, both men focused on drugs to the detriment of other school-related concerns.[12] Students certainly need a healthy learning environment, and one way of achieving it is to make the environment drug-free. But there are other problems of tantamount importance—such as structural inequalities, oversized classrooms, inadequate funding and resources—that were not part of their platforms.

The last and probably most disturbing aspect of this connection is that Clark's pronouncements about blacks become like putty in the hands of bigoted conservatives. Clark has been quoted as saying, "There's something wrong with my race, I must confess" (Kirp 1989, 38). He claims to be committed to assisting and uplifting African Americans, but statements like the following work against his uplifting measures: "I'm convinced that by the year 2000 we will self-destruct. Part of the reason for our precarious perch on Mount Oblivion is that Black men have faltered . . . there are some sorry Black men out there. I don't think any other race is as dependent on its women to carry the burden" (Chapelle 1989, 70). Clark evidently believes that he is offering insightful criticism and proposing a better plan of action, such as when he declares that "women unhappy at home—in their 40s and 50s, after menopause, when they're more consistent—can give untiringly to me of their services" (Kirp 1989, 39). However, his proclamations are often reactionary and detrimental to antiracist struggles, for they reinforce white racial hegemony and perpetuate the very ideologies that oppress the same people he purports to uplift. "The whites love me," according to Clark. The accuracy of his assertion is uncertain, but some conservatives undoubtedly appreciate having his rhetoric to appropriate for use as ammunition since it makes their work a lot easier. While on the subject of appropriation, it will be useful to explore how the filmmakers of *Lean on Me* sell their conservative messages through their filmic canonization of Clark.

Validating Joe Clark's Pedagogy

From the beginning of *Lean on Me*, it is clear that Clark will be the focal point of the film. The character is grandiose, commanding, driven, and aggressive, owing in part to Morgan Freeman's energetic performance. It is likely that the film's audience knew of Clark prior to viewing the film, for even if they had not witnessed the film's promotional trailers, they could have learned of him from his *Time* magazine cover story, seen him on ABC's *Nightline* or CBS's *60 Minutes*, or simply read of him in a local newspaper. Based on Clark's pre-movie popularity, then, audiences could have had a chance to come to their own conclusions about his educational philosophies and practices. Shortly after entering the theaters showing *Lean on Me*, however, viewers quickly became aware that the filmmakers were taking a supportive stance and were offering Clark as a hero.

Interestingly, though, the film does not only present Clark's "positive" side; he has, in fact, several traits that undermine his overall messianic image. The character is abrupt, accusatory, rude, and authoritarian. Furthermore, he appears megalomaniacal (a portrait not at odds with the real Clark) and in the opinion of many reviewers, fascist. "This is not a damn democracy," Clark yells to the faculty at Eastside in the scene of his first meeting with them. "We are in state of emergency, and my word is law!" Later in the film, Clark fires a music teacher in front of her class for what he sees as her lack of respect for his wishes (what he later calls "insubordination").

The inclusion of Clark's drastic and offensive actions may initially seem odd or contradictory in a film whose purpose is to heroicize him. One commentary suggests this by saying, "[T]he film poses some unusual problems: it makes Clark a hero while pointing to aspects of his character that trouble even those who support his goals and understand the harshness of his methods" (O'Brien 1989, 245). This is not so unusual, though, if one considers that narrative works of art often present the distasteful aspects of a character as long as the audience's sympathy can ultimately be won. O'Brien's explanation of the contradiction supports this hypothesis: "The film's faults are forgivable if you find Clark forgivable, which, on balance, I still do" (245). There is more to O'Brien's statement than perhaps he realized: that this was exactly the filmmakers' intention. In its representation of Clark, the screenplay is amazingly faithful to his real-life beliefs, values, and approaches. But the film's creators make other occasional changes in Clark's story and incorporate other narrative flourishes so that the revisionist account provided in *Lean on Me* presents such weak opposition to Clark and such substantial validation that the only position viewers are invited to take is to champion the champ.

The Politics of "Realness"

Lean on Me opens with the words, "The following is based on a true story" superimposed over a long shot of Eastside High, circa 1967. This text is presumably offered for the benefit of those unfamiliar with Clark's tale, lest they find the subsequent narrative too incredible to believe. This technique may also serve another purpose: by asserting the film's basis in "truth," it allows the filmmakers to escape the critique of viewers who might find the crux of the story unbelievable, while simultaneously allowing them some freedom to alter the narrative at will by denying that it is entirely factual.

The minds behind *Lean on Me* undoubtedly realized the usefulness of this trick. The producer said the following about the film: "We're not doing a documentary. We're doing a movie. And we're doing entertainment—and entertainment sometimes gets a little manipulative. I will opt to move the audience to laughter and tears with a little white lie anytime. Just because someone pays their $7 that doesn't mean they have to be told the truth" (Finke 1989). Ironically, Clark professed his faith in producer Twain's "honesty and integrity," and supported the film because he had a good rapport with Schiffer and Avildsen—little white lies and all (Finke 1989).

Clark's endorsement of the filmmakers' vision should not be surprising, given the consonance between his positionality and that of his character in the film and of the film itself. One would expect that a man who would manage a school in the same way as a prison (Stemer 1995) would sanction a text whose underlying message extols discipline above all else in the education of underprivileged blacks and Latinos. Armed with Clark's approval, the filmmakers were unimpeded in their attempt to present their version of the principal's story, a right-wing fantasy of one man taming hordes of "savage" students.

There are at least a few fabrications or white lies of consequence in *Lean on Me*. One of these is the fictionalizing of Clark's marital status. During an argument that Clark has with Superintendent Napier over the propriety and legality of his leadership tactics, Napier berates him for alienating people. "Look at you. You have no life, your wife left you . . . I ought to walk out on you myself." But at the time of the film, Clark was actually married with three children, two from his first marriage. Clark claims that the filmmakers felt they had to falsify his divorce, without offering him an explanation. "I didn't understand why," he said, "but I didn't even question that because it wasn't important. Knowing the media, it could have been worse" (Finke 1989). Why this falsehood? Perhaps the filmmakers wanted to paint the picture that he was unbearable to everyone, including his family members, or to establish the hero as the typical lone individual. Whatever their intention, this element reinforces the myth of the collapse of black family structure due to internal forces.[13]

A second embellishment is particularly germane to the one of the film's central tensions—the need to raise the test scores of Eastside's students. Before Clark's arrival at the school in the film, only 33 percent of the students passed New Jersey's minimum basic skills test. After their year of academic boot camp, an overwhelming majority of 85 percent pass the test. In actuality, the improvement in scores occurred, but not so drastically. The percentage of freshmen who passed the reading test went from 49 percent before Clark to 55 percent after his first year there, and the percentage of freshmen passers of the math test increased from 56 percent to 82 percent during the same time. This embellishment helped the filmmakers' to solidify Clark's savior image. And, since the results are announced at the end of the film, it serves as the ultimate recovery of Clark's image from public criticism and an implicit justification of his oppressive pedagogy.

Another of the film's myths is more consequential in terms of its racial statement. As I mentioned earlier, the film shows Clark during his initial stint at Eastside teaching a class of mostly white students, while in reality his students were mostly black. This change could have been made just to render more stark the contrast between the Eastside of his teaching days and Eastside after its decline. The not-so-innocent implication behind this narrative addition, however, is that the decline of the school is inseparable from the race and class of the students who inhabited it.

One last white lie that makes a similar racial statement is woven throughout the narrative and is relied upon for the denouement of the film. In order to prevent drug dealers from entering Eastside, Clark chained all of the school's doors and instructed that they be locked at all times. Clark refused to remove the chains even when he was alerted that they violated the fire code. He was eventually arrested and placed in jail for the infraction. As he sat in his cell and the school board decided whether or not to oust him from Eastside, a massive crowd of his students marched to the steps of City Hall to protest his imprisonment and to fight against his removal. As the African American female school board member responsible for his demise tries to quell the crowd, Clark emerges from the building, takes her bullhorn from her, and attempts to convince the students to disperse. At this moment, Clark's vice principal makes her way up to him to deliver the school's results from the state basic skills test. He reads the report and announces to the students the predictably stellar results. The mayor ultimately steps forward and displays his approval of Clark.

The real-life Clark did in fact "lock the school's 27 doors to keep out drug pushers, but he removed the chains after the city went to court to force him to do so. He was never arrested and never jailed" (Gellene 1989). The filmmakers obviously ignored this and used their fabricated version as the main element of tension in the film. The chained doors became a source of security for the students, the impetus behind a confrontation between Clark and the fire chief, and the motivation for the film's melodramatic climax. Clark, then, becomes a martyr for his beliefs. Instead of removing the chains at the fire chief's insistence, Clark remained firm in his decision to keep the "miscreants" locked out—but his choice of action also kept his students caged in like the wild animals or savages the film initially suggests they are. Even the school board attorney, Mr. Rosenberg, joshes Clark about his actions: "Thought you people didn't like chains?" Yet the joke falls on deaf ears, since Clark sees his way as the only solution, and his solution is celebrated through his exoneration at the ending. What is more is that the Eastside students cheer him on, both when he decides to chain the doors and when he is excused for it. This clever narrative twist reads like something dreamt up by Reagan in which an African American leader vows to "protect" his people by imprisoning them, the masses agree to this treatment, and the conservatives applaud the leader's bravery and conviction.

Leaning on Stereotypes

The right-wing pedagogy of *Lean on Me* operates not only through the falsification of Joe Clark's real-life experiences, but also through the use of other additions the filmmakers include in translating his story into its Hollywood equivalent. Several of these narrative elements certainly make the film, as the producer would have it, more entertaining. But some of them are extremely troubling in their reliance on stereotypes and reductionistic formulas, and detract from the film's potential for uplift.

The first example of this involves the creation of the character of Kaneesha, who is a bright, spirited African American female student at Eastside. When Clark first comes to the school, he recognizes Kaneesha in an assembly program as one of his former elementary school students. As the school year progresses, he keeps a watchful, caring eye out for her, and they develop a mutual fondness for each other. One day Clark finds her sulking outside his office and after a brief inquisition he finds out that she is afraid of being given away by her mother. Clark and Vice Principal Levias go to talk to Kaneesha's mother, who we soon learn is a single welfare mother in low-income housing and who regrets having become a mother at a young age. Clark's lecturing, Levias's concern, and Kaneesha's sadness convince the mother to overcome her embarrassment and pride and to struggle to make a better life for herself and her daughter.

Kaneesha quickly recaptures her vigor and good scholastic performance, and begins to date Richard, the class president. Soon afterwards, the inevitable happens: she becomes pregnant by Richard, who denies the allegation. Clark discovers her crying in the auditorium with a friend and learns of her pregnancy. Kaneesha tells him, "I never meant for this to happen." "You girls never do!" he responds. In one fell swoop, the film presents a troika of conservative stereotypes about blacks: the welfare mother, the pregnant teen, and the (potentially) absent father. And not only does the narrative simply offer these "types" to the audience, but it has Clark take an almost condemnatory stance in relation to them. When white students in 1980s school films become pregnant or are teen mothers, as in *Teachers*, they are shown sympathy from an authority figure and never consider dropping out of school; when the students are black—here and in *The Principal*—they immediately plan to leave school and/or are interrogated and lectured about their choices.

Another example is linked to the issue of the chained doors. In the film, Leona Barrett is the black mother of one of the "miscreants" that Clark expelled from Eastside immediately after his arrival. Incensed at Clark's action, she speaks out against him at a meeting of concerned parents. Unfortunately for her, she is outnumbered by those parents willing to give him a chance, so she has to pursue other means of attacking Clark. Barrett pleads with the mayor to have Clark arrested for violating the fire code, and eventually bargains with him to give her a seat on the school board in exchange for her support in the upcoming election. She ends up in cahoots with Mayor North and Fire Chief Gaines (who turns out to be a bigot), and it is due to her insistence that Clark is finally arrested and jailed.

This series of developments seems to be crucial to the filmmakers' conservative project for a number of reasons.

For one, it can be read as an manifestation of the oft-expressed belief by conservatives that intraracial disputes are the obstacles to progress for nonwhites. Relatedly, the narrative depicts the betrayal of Clark by a black woman, not in the interest of justice but of vindictiveness, further suggesting that the enemy is not outside the race but within. Third, Barrett is set up as the angry black woman, implying that the audience should not buy into her criticisms of Clark, but rather should distrust her motives because they are driven by ambition, revenge, and anger. Last, by portraying all of Clark's opponents as opportunistic and distasteful, the film leaves little room for disagreement with or even reflection upon the central character's pedagogy and recovers him from the machinations of his enemies. Given that the film is primarily about the celebration of his accomplishments, the viewer is by extension discouraged from disagreeing with the pedagogy of the film itself.

A Pedagogy of Hope in Lean on Me*?*

Despite the conservative ideologies imbedded in the narrative of *Lean on Me*—such as the ignoring of structural forms of oppression, the stereotypical portrayals of nonwhites and the working class, and the explicit link between certain races and criminality—are there any progressive elements at work? Some are convinced the film has potential to be uplifting. One reviewer, for example, argues that regardless of those who see the film as an uncritical endorsement of Joe Clark, "[M]ost viewers . . . will find *Lean on Me* a satisfying tale of a man who does what is necessary" (Marcazzo 1990, 210). This argument proposes that the objectionable aspects of Clark's pedagogy are merely the unavoidable means to more noble ends. Hence, sympathizers who believe in the ends will be inspired by the redeeming qualities of Clark and of the film.

One can find at least a few seemingly inspirational elements of the film. What is problematic, though, is that most of them are counterbalanced by other effects. One example of this occurs in the scene in which Clark fires his music teacher for "insubordination." He dictates that she refrain from rehearsing Mozart with her students as they prepare for an upcoming concert, and instead demands that she teach them the school's alma mater. Clark's request could be interpreted as a challenge to the supposed supremacy of the Western canon and a subtle suggestion that equal emphasis be placed on the pride-instilling act of learning the school song. However, this interpretation overlooks the inappropriateness of the disrespect Clark shows the teacher, as well as ignores the pleasure students may have derived from showcasing their talent at a concert.

There are still other sources of contradictory moments of progressive and conservative pedagogies. In his rousing address to his students just before they sit for the state basic skills test, Clark encourages them by telling them that outsiders speak disparagingly about them and have little faith in their ability. Elsewhere in the film, he tells Napier that "we are being crucified by a process that is turning blacks into a permanent underclass." In these two instances the film offers a fitting indictment of the forces that oppress underprivileged youths. But the personal motto that Clark shares with the students early on flies in the face of his other criticisms. "If you do not succeed in life," he states, "I don't want you to blame your parents. I don't want you to blame the white man. I want you to blame yourselves. The responsibility is yours!" This conservative mantra of the responsibility of the individual isolated from society undermines the impact of Clark's more inspiring rhetoric. It also marks the previously noted trend in school movies away from providing a social context for students' behavior, leaving racial minorities to blame themselves. Clark's perspective is unlike that of Marcus Garvey, who has been lauded by conservatives for his self-help philosophy (Wright 2002). Garvey, as Lott (1992) smartly observes, outrightly acknowledged that social equality could never be achieved under America's "apartheid." Garvey, unlike Joe Clark, realized that bootstrap uplift had to be combined with a "denunciation of America's racism" (77).

Another example in *Lean on Me* of the juxtaposition of the progressive and the regressive manifests itself in Clark's decision to place chains on the school's doors. On one hand, through his decision to chain the broken doors to lock out intruders, especially when warned against the action, he displays conviction and concern for the welfare of his students. Clark's willingness to challenge the system to protect his students could be seen as an admirable quality of a leader, further highlighted by his acceptance of his just punishment for breaking the law. On the other hand, Clark could have provided leadership in another way: through negotiation. In the film, he bemoaned not having alarm doors "like the white schools have." The script could have had him bargain with the mayor or at least the superintendent to remove the chains once the broken doors were repaired or alarms were installed. He could have even sought the support of the Eastside parents given the considerable leverage he had with them. Instead of finding another solution to the situation, the filmmakers evidently chose to create the arrest–imprisonment–redemption sequence because it was better suited to their aesthetic and political ends.

In his review of *Lean on Me*, O'Brien writes that the film's "aesthetic sins are slight compared to the good the film may do for its intended audience. Despite its defects—even despite the defects of the personality at its center—the film dramatizes the need to nurture hope in inner-city schools" (1989, 245). O'Brien's emphasis on the need for schools to operate on the basis of a pedagogy of hope is a worthwhile point. Maybe the filmmakers agree with his position, which might explain their decision to exaggerate Eastside's performance on the state's standardized exam, or their choice to have the students' beloved principal imprisoned, only to have him be released and delivered back to Eastside to work his magical tough love. To the extent that viewers find these elements inspirational and meaningful is a testament that the film, in a sense, *worked*. But what this view does not account for is that the ideologies behind the film's "aesthetic sins"—most of which are racially discriminatory, patriarchal, and classist—are oppressive, and their oppressiveness overpowers any hope that the film teaches us to have.[14]

A Reaganite Fantasy

One commentator has said that "*Lean on Me* was not made for film critics; it was made to reach the kind of adolescents Joe Clark was able to reclaim for society" (O'Brien 1989, 245). Without disputing the fact that no cultural product is designed for all audiences, I should hope that a sustained analysis of *Lean on Me* can achieve the goal of exposing the conservative ideology of the film and help others to understand the film in this light. One has to wonder if O'Brien's statement would be made if Joe Clark's students had been mostly white and middle- or upper-class. Another educator, George McKenna, raised that exact question, and claimed that Clark "would not be tolerated if the students were not poor and black" (Scott 1988, 38).

McKenna, the erstwhile principal of George Washington Preparatory High School in Los Angeles, once found himself leading a similar constituency of students. Though not as well known as Clark, he was also the recipient of praise from Ronald Reagan and the subject of a biographical movie in 1986 (*The George McKenna Story*), albeit a made-for-television version. Over a nine-year period, McKenna transformed Washington Prep behaviorally and academically, through his commitment to nonviolence and love. McKenna exhibited concern for individual differences, needs, and problems without ignoring the institutional causes that are often responsible for their problems, a lesson conservatives refuse to learn. Yet his leadership was not without its critics, and his story not without its exaggerations; in other words, McKenna, like Clark, is not perfect. But his story does offer another model of somewhat successful leadership that was ripe for translation onto the big screen—and there are countless others indeed.

One of these countless others is James E. Watson, who was in 1989 the headmaster of Madison Park-Humphrey Center High School in Boston. Watson's school was similar to Joe Clark's Eastside High, in that both are urban institutions with more than 2,000 students who are predominantly African American and Hispanic. Both schools also suffered from political and physical neglect and were homes to low standardized test scores. Yet Watson's approach to dealing with the problems of his school was quite the opposite of Clark's approach in *Lean on Me*. Watson, who was described as a "very democratic leader" by one of Madison's assistant headmasters (Doten 1989), has indicated his disapproval of the flattering filmic depiction of Clark's pedagogy. In his school, Watson, unlike Clark, focuses on the necessity of working as a team "not only with staff members but with students, parents and the community" (Doten 1989). By his own admission, Watson's record of achievements is not unblemished; however, he has accomplished much at Madison-Humphrey High of which to be proud (Doten 1989).

So, why was Joe Clark's story privileged over all other hero tales, African American or not? It was not chosen only because of its controversial nature, for a character can arouse controversy without being egomaniacal or carrying a bat. Moreover, there are enough similarities between the histories of Clark and McKenna or Watson to have made the latter ones easy choices for the filmmakers to exploit. There is even a possibility that McKenna would have supported the desire of *Lean on Me*'s creators to have their film revolve around a sense of hope. In fact, he suggested as much in his defense of the flattering portrait of his story in *The George McKenna Story* by saying, "Schools like [Washington Prep] need to be uplifted to a level of reverence to show that they can work" (Kirp 1989, 43).

In spite of all this, McKenna's story, Watson's, or another one similar to them would have been rejected for at least one reason: its incompatibility with the conservative agenda that the filmmakers of *Lean on Me* apparently wanted to promote. While Watson and McKenna have argued against Clark's methods and criticized the ideologies of the film, Clark has made assertions like "There's something wrong with my race, I must confess." In an era that featured Reagan's traditionalism, Bennett's coded racist discourse, economic malaise, and anti–civil rights policies and sentiments, Clark's beliefs were the pieces needed to complete the puzzle of conservative pedagogy and feel-good politics. In their attempt to create a tale of uplift and inspiration, the filmmakers of *Lean on Me* ended up creating what looks more like a right-wing dream sequence. At the end of this dream, blacks are singing in unison, fulfilling the performative role, a role that has always been pleasing to Hollywood (Bogle 1973, 118). And Joe Clark comes forth, ready to lead them through the chained pearly gates to the educational promised land, bat and bullhorn in tow.

4

Isolating and Protecting Whiteness

George Herbert Walker Bush was inaugurated as the 41st President of the United States of America on January 20, 1989. His election several weeks earlier was not easily achieved, but given the state of affairs after Reagan's reign, more than easy solutions were needed to deal with the country's uneasiness about its future. Through much finagling, Bush claimed the presidency over the embattled Democratic candidate, Michael Dukakis. Bush already had a base of elite supporters whose upbringing and socioeconomic status resembled his own; however, winning the 1988 election would require a broader base of constituents. So the Bush campaigners garnered the support of an odd alliance of voters: in addition to the wealthiest classes, they attracted the right wing (including the "Old Right," "New Right," and "Christian Right"), the so-called Reagan Democrats, and the suburban independents (Duffy and Goodgame 1992; Moen and Palmer 1992). Of course, in order to appeal to the diverse beliefs of these groups, Bush had to accommodate their various interests. As if these were not enough platforms to espouse, Bush was also faced with the challenge of how to retain the momentum of the Reagan administration and take credit for its successes while rejecting the unappealing elements of the Reagan legacy.

Bush's rhetorical juggling led many observers to comment on his lack of a "mandate" or agenda and on his lack of consistency. Duffy and Goodgame (1992) point to cartoonist Garry ("Doonesbury") Trudeau's introduction of "Skippy,"

Bush's evil twin. "Even Bush's advisers," they write, "were somewhat embarrassed to watch as the vice president alternated each day from Dr. Jekyll to Mr. Hyde. 'We were totally schizo,' said one top campaign official" (30–31). Yet Bush emerged as the victor over Dukakis—but not without a close and sometimes nasty struggle on Bush's part. As soon as the election was over, President Bush wanted to empty the campaign trash and begin serving with a clean slate. He tried to exonerate himself from the malice, indiscretion, and "schizophrenia" of his campaign by claiming that "anything goes" in politicking, but that his leadership would be a different story (Duffy and Goodgame 1992, 34–35).

Leadership, Bush Style

On the basis of what little could be discerned about Bush's presidential agenda prior to his election, most Americans believed that he would be a more moderate leader than Ronald Reagan, and his "kinder, gentler" slogan certainly reinforced this impression. Now, after the end of his term in office, there has been some evaluation of what his real political convictions were. Shortly before Bush left office, Duffy and Goodgame (1992) had determined from the statements and actions he made before and during his tenure that the former president held "a deeply conservative world view" (64). In terms of his attitude toward change, established institutions, hierarchies, and the role of government, they argue that Bush's conservatism exceeded that of Reagan. Bush, the authors suggest, was the ultimate defender of the status quo (Duffy and Goodgame 1992), and it is this political approach that has led many pundits to consider Bush's presidency as simply a guardianship of the Reagan legacy.

Malecha and Reagan (1992) seem to be in agreement with at least part of the above contentions. Contrary to *Time* magazine's January 7, 1991 depiction of Bush having one domestic personality and a different foreign one, they contend that in domestic affairs and foreign affairs, Bush has approached relations with Congress in a predictable way. One can view Bush as predictable—instead of as a leader with dual personalities—by recognizing his deep commitment to hierarchy. In this way, their evaluation of Bush resonates with the contention by Duffy and Goodgame (1992) that Bush is indeed a man of solid convictions.

The problem with Mr. Bush's leadership, then, was not his lack of conviction as much as it was his lack of consistency and sustained effort to see his major campaign commitments realized. Duffy and Goodgame (1992), for example, assert the following: "Bush often reveals his lack of interest and cynicism toward a domestic issue by raising it for a transitory political purpose, then failing to follow through" (97). Thompson and Scavo (1992) make a similar point in writing that "even in the domestic policy areas Bush claims to see as critical, there has been little personal involvement or follow-through from which a domestic vision could be developed" (161–162). They grant Bush the benefit of the doubt by acknowledging that his political philosophy and approach may have precluded a broader or more aggressive domestic agenda. Nevertheless, after accounting for all of the factors affecting his leadership, they (and many others) still found that the domestic agenda got short shrift.

Consider his actions on particular issues. Despite his attempt to leave the dirty past of the campaign behind him, there were certain campaign pledges that Bush intended and pretended to honor. As a presidential candidate, Bush made education a focal point of his campaign, claiming to want to be the "education president" and talking of high American standards and quality schools. Once in office, he promoted an educational agenda under the name of "America 2000" that was by most standards an ambitious plan that would have been difficult to fund. Yet Bush ultimately failed to contribute sufficient effort toward achieving his aims (Duffy and Goodgame 1992; Thompson and Scavo 1992). It was not until April of 1991 that Bush formally adopted the national goals, a long period of time for a president who emphasized the priority he would place on education. Furthermore, he did not uphold his campaign promise to fully fund the well-respected Head Start program (Duffy and Goodgame 1992). Neither did Bush stand behind his plan for federal tuition vouchers for public school students to redeem at private schools of their choice—at least not until it was time for him to make reelection campaign promises in 1992 (Greider 1992). Bush was consistent, however, in his maintaining of the status quo by carrying the Reagan banner of school choice into the next decade.

President Bush's approach to drug control, another of his campaign platforms, suffered from the same malady as his approach to his educational agenda. His dealings with the drug problem might be characterized much like his handling of the education issue as described by Thomson and Scavo (1992): "Bush's statement of broad goals and ambitious schedules has been accompanied by a lack of consistent follow-through both by himself and those in his administration charged with pursuing his policy goals" (156). Bush also upheld the Reagan legacy in his drug control strategy, thanks in part to the involvement of Reagan supporter William Bennett. Most of his campaign promises concerning the "drug war" were simply repackaged Reagan proposals (Lusane and Desmond 1991; Thompson and Scavo 1992). Although Bush departed from these promises to some degree after he was elected, the new directions he followed resulted in only "incremental changes in Reagan policies" (Thompson and Scavo 1992, 154). In these and other areas, President Bush failed to follow through on his own agenda while consistently perpetuating the Reagan influence.

Bush on Racial Equality

Duffy and Goodgame (1992) contend that "when he does follow through on a domestic issue, Bush often adopts a carefully worked-out straddle," and that the biggest straddle of his presidency came on civil rights (99–100). Shull (1993) is in agreement on Bush's "middle ground" positionality on civil rights, but he goes further in declaring that Bush issued strong statements and actions that were unexpectedly more conservative that those of Ronald Reagan. In contrast, Mullins and Wildavsky (1992) suggest that the contradictions that have been highlighted reflect his attempts to reconcile his individualist and hierarchical inclinations. While there are varying opinions on Bush's handling of civil rights issues (more to be discussed below), most seem to agree about his unwillingness to maintain a consistent position. Jones (1992), for instance, evaluates Bush's civil rights record shortly before he left office on the basis of his policymaking and his appointments. As for the civil rights policymaking, Jones (1992) examines his support in three areas: school desegregation, affirmative action, and voting rights. The author concludes that Bush was "kinder and gentler" than Reagan on the three issues, but wavered in his support for the first two issues and was ambivalent toward civil rights in general.

Bush's record on the political figures he chose to endorse and appoint appears a bit more problematic, occasionally leading to accusations of race baiting. Jones (1992) remarks that "Bush's appointments to key administrative and judicial posts have also been quite mixed," some actively supportive of civil rights, others fervently against it (185). Jones cites Bush's selections of Arthur Fletcher as chairman of the CRC and Louis Sullivan as Secretary of Health and Human Services as examples of those seen to be favorable to civil rights. It is likely that Bush also gained the confidence of some civil rights groups through his appointment of Colin Powell. On the other hand, Assistant Attorney General for Civil Rights William Lucas and Supreme Court Justice David Souter were viewed, respectively, as unqualified and unconcerned, and civil rights groups received those appointments as slaps in the face from a president who had professed his commitment toward racial equality in America (Jones 1992). According to Duffy and Goodgame (1992), another controversial recipient of Bush's support was North Carolina Senator Jesse Helms, for whom Bush campaigned heavily in 1990 against Harvey Gantt, the African American Democrat candidate. This was a sign of the president's "unprincipled approach to the civil rights debate," particularly given Helms's racially coded television campaign ads and reported intimidation of black voters (101–102). One other Bush appointee that cast doubt on Bush's commitment to racial equality was Michael E. Williams. While he was the civil rights officer in the Department of Education, Williams—who is African American—proposed a halt in federal funding for colleges and universities that awarded scholarships based on race. His proposal elicited forceful opposition, and the Bush administration opted to distance itself from Williams's plan.

Among Bush's moves in the appointment arena, the ugliest episode of racial politics involved the Senate hearings and subsequent nomination of Clarence Thomas to the U.S. Supreme Court. Thomas had been nominated by President Bush in 1991, presumably to symbolically replace the African American presence on the court left vacant by Thurgood Marshall's retirement. Bush's nomination was also seen as an effort to appeal to the diverse civil rights interests of American citizens, again demonstrating what Duffy and Goodgame (1992) called his "carefully worked-out straddle." In this instance, however, Bush's straddle turned out for many reasons to be more divisive than appeasing. In light of Thomas's expressed beliefs and judicial history concerning affirmative action, abortion, and other issues, many critics accused Bush of intentionally inciting battles within African American, feminist, Democrat, and left-wing camps. The level of indignation of the Bush critics was further elevated by the shameful (and shameless) Senate confirmation hearings, which partly involved Thomas's former employee Anita Hill and her charge of sexual harassment against Thomas (Morrison 1992; Shull 1993).[1]

Did George Bush simply maintain Reagan's civil rights legacy, or did he pursue a qualitatively different agenda? Shull's (1993) answer to that question is that "Bush actually went beyond Ronald Reagan in his efforts to cut back the federal role in civil rights" (4). In his actions and rhetoric—including his reprehensible capitalization on the Willie Horton election campaign ads and his partial blaming of the 1992 Los Angeles uprising on the social programs of the 1960s and 1970s—Bush "played the race card" in tricky and underhanded ways. Perhaps Shull is in the minority of those who would place Bush to the right of Reagan on civil rights issues, but all of the aforementioned writers tend to agree on two points: that President Bush walked a narrow, unsteady tightrope in the civil rights arena, and that he often maintained Reagan's conservative legacy.

Culture Wars and the Academy

Ronald Reagan used most of the 1980s sowing and cultivating the seeds of a reactionary racial politics, with George Bush acting as his chief horticulturist. Bush, of course, would replace Reagan and as the above discussion indicates continue the legacy. The fruits of their labor matched similar efforts among the American populace in the economic, educational, and cultural (and their overlapping) fields. One of these fields of intersection involved the role and influence of Western civilization in American educational and cultural institutions. The Western canon of figures, texts, traditions, and values lay (and continues to lie) at the center of a highly contentious debate over what knowledge should be learned, why and how it should be taught, and in what doses. In general, supporters and critics believed that the supremacy of the Western canon in America was threatened, but they believed it for different reasons.

In one camp were mainly conservatives and neoconservatives who feared that the status of Western civilization in colleges and universities, high schools, and other cultural milieus was under attack and in danger of being subverted by militant radicals. The biggest—though certainly not the first—spark for the debate fire was Allan Bloom, although this was probably unintentional on his part. Professor Bloom, who was known for his scholarship on Plato, Rousseau, and Shakespeare, became famous and infamous for his 1987 text—*The Closing of the American Mind: How Higher Education Has Failed Democracy and Impoverished the Souls of Today's Students*—which commanded significant attention both inside and outside the academy and sold millions of copies. In situating the book in his sweeping intellectual history of the twentieth century, Peter Watson contends that it had "set the cat among the pigeons in the academic world" and then broke out of the "scholarly ghetto for which it had been intended" (2001, 721).

Bloom, in his "meditation on the state of our souls, particularly those of the young, and their education" (1987, 19) reflects on a myriad of subjects including nihilism, sex, love and eros, race, music, and the sixties, mainly in the context of the American university and in relation to Greek philosophy. More pertinently, in several places he offers his perspective on the canon wars. In "The Student and the University," for example, he writes, "I am perfectly aware of, and actually agree with, the objections to the Great Books cult," then enumerates a number of those objections. "But," he continues, "one thing is certain: wherever the Great Books make up a central part of the curriculum, the students are excited and satisfied, feel they are doing something that is independent and fulfilling, getting something from the university they cannot get elsewhere." Thus, while expressing agreement with the critiques of the "Great Books cult," Bloom implicitly celebrates their nonpareil appeal and explicitly defends their "judicious use" (344). Here and throughout his book, Bloom laments the "crisis of liberal education" and offers an extended justification for the role of the Great Books in addressing this crisis.

The state of education in our nation's high schools caused similar worries over the displacement of certain forms of knowledge privileged within the Western canon. Other than Bloom, E. D. Hirsch, Jr. is easily recognizable for his attempt to address this perceived displacement. Hirsch was apparently concerned over what was lacking in American secondary education, and, in 1987, he addressed these concerns in *Cultural Literacy: What Every American Needs to Know*. In his work, Hirsch dares to delineate the base of knowledge every citizen of the United States should possess in order to be "culturally literate." He argues that cultural literacy "constitutes the only sure avenue of opportunity for disadvantaged children" and provides the "basic information needed to thrive in the modern world" (1987, xiii). Hirsch's attempt to deal with widespread illiteracy and the failures of American schools seems noble enough, but his work ends up being little more than a top-down prescription (although he claims otherwise) for its "chief beneficiaries . . . disadvantaged children" (xiv). Two of the book's most serious problems are that it fails to question who has the power to define what cultural knowledge is important, and that it ignores the destructive aspects of the history of European colonization and its remnants in the "shared culture" of the United States of America.

Others, like Arthur Schlesinger, have weighed in on the matter. Schlesinger (1991) argues that the spread of the "cult of ethnicity" will lead to a "disuniting" of the United States of America unless its effects are counteracted. His prescriptions for countering this divisive force are assertions of the supremacy of Western ideals and barely camouflaged attacks on Afrocentrists. And we clearly cannot forget one other central figure in the "defense of the West"—William Bennett. In yet another instance, Bennett has been a visible, vocal, and influential player in the racial politics of the Reagan–Bush legacy and the right wing in general. As Chairman of the National Endowment of the Humanities, Secretary of Education, and Director of Drug Policy under Reagan and Bush, Bennett has had undue influence on the political and cultural institutions of the United States. He has spoken and written widely on what he sees as the threatened state of our "common" culture, language, values, and institutions. For example, in his collection of speeches delivered while he was Secretary of ED, Bennett (1988) raises in one of the speeches the question, "Why Western Civilization?" He answers it in part by lauding the Western tradition as an "unparalleled resource" (197), albeit without demonstrating much knowledge of any other traditions or civilizations. He concludes, like his ideological counterparts, by declaring Western ideas to be the "last, best hope on earth" (201).

Despite any claims to the contrary (e.g., Bloom 1987, 22), the tone of the writings of Bloom, Hirsch, Bennett, and the others is overwhelmingly nostalgic, which Blumenthal (1988) has observed as also being characteristic of the Reagan aesthetic. Like President Reagan, Bush, Bloom, and Bennett perceive the sixties as the beginning of the end and yearn to recover the greatness of the past. Bush, for example, blamed the 1992 Los Angeles uprising in part on social programs of the 1960s and 1970s (Bush 1993). Bennett (1988) claims that the quality of American education was the unfortunate victim of the "assault on moral and intellectual standards" during the sixties and seventies (9). For his part, Bloom writes, "[s]o far as universities are concerned, I know of *nothing* positive coming from that period; it was an unmitigated disaster for them" (320; emphasis added). Sentiments like these undergird the following observation from Blumenthal (1988): "Conservatives of the Reagan era often blamed the sensation of vertigo they felt in a pluralistic America on moral relativism" (260–261).

Such was the state of things in the United States in the spring of 1989. There was Bush's continuation of the conservative Reagan legacy; the strident defense of the West by those scared into battle in part by the release of *A Nation at Risk*; and the continuing abandonment of the inner cities by middle-class whites discussed in chapter 3. All of these elements formed a richly textured backdrop against which appeared an intriguing new school movie: *Dead Poets Society*. In the few years before, the casts of the most popular Hollywood school films had become increasingly multiracial. White students, however, had not been totally abandoned; they had simply gained some distance on screen from their nonwhite, urban counterparts, imitating current demographic patterns.

The Preservation of Tradition: Unearthing *Dead* (White Male) *Poets*

Like Bush's educational agenda during his term in office, the period after *Lean on Me* was limited in scope and size for school films. Hollywood did manage, though, to produce during Bush's presidency one very popular, profitable, and significant school film. In June of 1989, Disney's Touchstone Pictures released a weighty drama entitled *Dead Poets Society*, which seems an unusual choice given the film industry's tendency to distribute lighthearted comedies and action-adventure movies during the summer months. But the film did have something important in its favor: actor Robin Williams, who had starred in Disney's *Good Morning, Vietnam* (1987) two years before, would be playing the lead character. Williams's unique style of performing, along with Disney's reputation and the film's PG-rating, undoubtedly worked to attract viewers who otherwise might have been repelled by the title or subject matter of *Dead Poets Society*.

However, unlike the chaotic urban setting of school movie *Lean on Me*, which had appeared in movie theaters only a few months earlier, the filmmakers of *Dead Poets Society* chose a more idyllic locale for its backdrop. School films of the previous years—such as *Teachers, The Principal, Stand and Deliver*, and *Lean on Me*—featured inner-city high schools with multiracial casts (the more recent the film, the fewer the number of white students). *Dead Poets Society* departed from such a trend: the setting of its educational institution was far from New York, Los Angeles, or any such metropolis; its homogeneous cast was quite unlike the non-white faces school film audiences had become accustomed to seeing; additionally, the film is set in the past instead of the present. Equally important is the fact that the film's messages are in many ways vastly different from those of its predecessors.

Dead Poets Society primarily concerns the arrival of former Welton Academy student John Keating (Robin Williams) as an eccentric, new, English teacher and the impact he has on his students. Nothing seems out of the ordinary about Keating's background, given that he graduated from Welton with high honors, went on to become a Rhodes scholar, and subsequently taught at a highly regarded school in London before returning to teach at his alma mater. In contrast to the other staid, traditional instructors at Welton, however, Keating immediately influences his students with his unusual approach to teaching literature and disrupts, for better and for worse, the usual functioning of the institution.

Yet the focus of *Dead Poets Society* is not Keating. Instead, the movie is about Keating's pedagogy and how his students make use of it in their lives inside and outside the classroom.

Keating's motto is "*Carpe diem* . . . seize the day. Make your lives extraordinary," and since he also teaches his students that "words and ideas can change the world," it is largely through language and other symbolic behavior that he encourages them to be extraordinary. Thanks to Keating's impassioned instruction and unorthodox use of poetry, the students are inspired by him and take his teachings to heart. Many—but not all—of his students are so transformed by his challenge to make their lives extraordinary that they take his mottoes out of the realm of the classroom and act on them. One of these "acts" is the resurrection of the Dead Poets Society, an underground group of poetry enthusiasts of which Keating was a founding member. The film focuses on the seven disciples who revive the society and follows their attempts to "seize the day."

Two of the central characters, Neil Perry and Todd Anderson, are roommates. Perry and Anderson are an interesting pair for a number of reasons: generally speaking, their lives are in loose opposition to one another. Perry is a popular student with an established base of friends; Anderson, on the other hand, is a new student who is begrudgingly at Welton and lives in the shadow of his older brother, a past valedictorian at the school. Perry is extroverted and energetic, while Anderson is awkward, quiet, and not very self-confident. Anderson suggests that his parents are too self-absorbed to show interest in him, whereas Perry's dictatorial father exerts inordinate influence over his son's life. Lastly, Anderson is the progeny of the upper-crust parentage that is typical of Welton; however, Perry's family is not as affluent as the rest of the students' families. Nonetheless, the two boys grow to like and trust each other as the story progresses.

The film also closely follows five classmates of Perry and Anderson. One of these students is jokester Charlie Dalton, who, the film suggests, comes from the wealthiest family of all. Dalton is the most zealous about Keating's charge to do something great or unusual in order to add spice to life. Halfway through the film, Dalton adopts a new identity—signaled by his adoption of a new, ethnic name—which he keeps even as he is suspended from Welton for his pranks.

Another student is Knox Overstreet, who is inspired romantically by Keating's use of poetry. His inspiration gives him the courage to pursue a young woman from the local public high school in spite of her jock boyfriend.

The three other lesser-followed classmates (referred to by last name) are: Richard Cameron, a punctilious student who is never fully swayed by the dynamism of Keating and who is always aware of the consequences of the group's transgressions; Gerard Pitts, a gangling electronics buff who is initially reluctant to engage in the group's risky endeavors but is ultimately liberated by them; and the bespectacled Steven Meeks, who is portrayed as a smart student who internalizes Keating's teachings without much prodding. All seven of the students are profoundly affected by Keating, but they embody his pedagogy in different ways and to varying extents.

"Divergent" Views about the Film

Around the time of the release of *Dead Poets Society*, there was some discussion in the popular press about the film being a semiautobiographical account of the experiences of the author of the screenplay, Tom Schulman. Seidenberg (1989), for example, writes that "Schulman modeled Keating on several teachers who had a lasting effect on him, including theater guru Harold Clurman," who taught Schulman at the Actors and Directors Lab in Los Angeles (57). An article by Farley (1989) tells a somewhat related story about the film's limited basis in reality. It claims that the movie is not autobiographical, except for the inspiration Schulman received from an unnamed acting teacher in Hollywood. This is "the only strong personal experience Schulman inserted into '*Dead Poets*,' apart from the love of poetry" partially aroused by his physician-father's belief in the wisdom revealed in it.

However, Pitts (1990) offers a convincingly different, albeit uncorroborated, perspective on the origin of the film's narrative. Pitts claims to have been a 1965 classmate of Schulman and a student of Sam Pickering—"the real Mr. Keating"— at Montgomery Bell Academy in Nashville, Tennessee. Although

Schulman did attend the school, he asserts that it "wasn't really like the school in the movie" (Farley 1989). Pitts (1990) tells a different story:

> The principal characters are based on real people, and many experiences actually happened in some form. There was no suicide, however, and no Dead Poets Society. The adults in our lives were not all ogres; in fact, they were pretty much like we are now. (3)

Pitts refers to Mr. Pickering as a "very eccentric, brilliant character" who had an extraordinary effect on him and his classmates, and goes on to describe a person who closely resembles the Keating character. He recounts incidents like those in the film, like Pickering making proclamations from his desktop, requiring them to recite quotes before kicking the ball during soccer practice, and exhorting them to "seize the moment, for tomorrow may be too late!"

Pitts (1990) points to other elements from the movie that recall Pickering's 1965 class. "I know from talking to [Schulman]," writes Pitts (1990), "that he kept all his books and notes from Pickering's class, which he used in the film . . . The literary passages in the film were ones Tom underlined in class discussions" (3). He also identifies characters in *Dead Poets Society* whose names are similar to those of his and Schulman's classmates, including the movie's Gerard Pitts, resembling his full name, Greenfield Pitts. Emphasizing another strong similarity to the film, Pitts (1990) remembers how Pickering illustrated through literature a constant message: "Think for yourself; make your own decisions; be responsible for your actions" (6).

Besides the articles already mentioned, there are numerous reviews on *Dead Poets Society*—by film critics, teachers, students, and others—and the reviews express various interpretations of the movie's messages. The large majority of literature written about *Dead Poets Society* consists of popular press reviews. Many of the reviews are glowingly favorable, or at least generally complimentary with particular reservations. For example, Dawes's (1989) review describes

Dead Poets Society as a "lyrical, gorgeous and tough evocation of the unsteady fires of youth" (26). Siskel (1989) writes that the movie is "refreshing if obvious," and picked it as his "Flick of the Week." Wilmington (1989) calls for respect and affection for the film despite its flaws, and Howe (1989) sees the film as being "solid, smart entertainment." Additionally, all of these reviewers are in general agreement that the film takes a stand against conformity and invites viewers to question traditionalism.

Other reviews of the film consist of a mixture of favorable and unfavorable comments or they focus on the effectiveness of one or another element, this element often being Robin Williams's portrayal of Keating. Other topics of discussion are the performances of the actors playing Keating's students, Perry's father, and Mr. Nolan, Welton's headmaster; the film's cinematography; and the creativity of the screenplay. There are also reviews that take a more critical stance on the movie. In opposition to those who highlight the film's subtle rejection of conformism, columnist Dave Kehr (1989) critiques the film's "theme of forced individualism." Another take on the movie, from the perspective of an Australian educator, doubts the ability of Keating's teaching to change real students' lives for the better: "Though the film's poetry and music are unquestionably the food of love, the simplistic dichotomy of institutional authority and student freedom sheds little light on the complex reality of private and public schools in 1989" (Collins 1989, 75). From these few examples, one can understand the diversity of printed opinions about *Dead Poets Society*. Yet it is the divergence of these readings that will underscore the film's resonances with the conservative ideologies of Bush, Bennett, Bloom, and, broadly speaking, the Reagan legacy.

Ambivalence about Tradition

Interestingly, it seems that there is evidence in *Dead Poets Society* to support the wildly diverse perspectives. On one hand, a fair characterization of the narrative has to make mention of the film's exploration of the desire for individual freedom in the face of institutional pressures to conform. The film in fact appears to take a position in favor of finding one's own voice and going against the grain. Indeed, the poems selected in the film are intended to encourage the students to follow the road "less traveled by."[2] And even though Keating is ultimately punished by the Welton establishment for his teaching, the movie's climax vindicates Keating, and his pedagogy lives on in the decisions made by his students: Overstreet woos his love interest with poetry and wins her over; Dalton, though expelled, keeps his alter ego alive to the end; Anderson stands on his desk at the conclusion in a show of respect for Keating, flagrantly disobeying the headmaster, and several others join him; and even Perry's suicide could be seen as an act of defiance of his father and a display of individual freedom inspired by Keating.

On the other hand, the filmmakers of *Dead Poets Society* appear to have been unsure about their commitment to the idea of promoting independent thinking, extraordinary acts, and unorthodox behavior. At one point in the movie, Charlie Dalton argues in the school journal on behalf of the Dead Poets Society for the admission of women into Welton (although his motives are far from altruistic). This uproar prompts an assembly of all Welton students in order to flush out the responsible party. During the assembly, Dalton pulls an outlandish prank for which he is paddled by the headmaster. While he is recounting the details of the punishment to the other students, Keating enters and reprimands him for performing such a foolish prank. Dalton is startled by Keating's call for caution, until Keating explains that "sucking the marrow out of life doesn't mean choking on the bone." Keating then makes his exit, but not without an ambiguously humorous remark about the prank.

A similar uncertainty surrounds Perry's suicide. Perry appears to have taken seriously Keating's appeal to enjoy life while we can before we become "food for worms." Perry made his life "extraordinary" by disobeying his father in order to star in a play, but to what end? As with Dalton, the film underscores the fact that one must take responsibility for one's own decisions, but having Perry removed from Welton by his father and then commit suicide seems a high price to pay for his desire to be great. Again, as with Dalton, his decision was not *totally* his own—it was partially inspired by Keating's pedagogy. Thus, in some cases the students are rewarded for unorthodox behavior, and in other cases they are punished. Do these examples point to fundamental inconsistencies in the film's message?

Perhaps the film's mixed messages are so salient because they are unusual for Hollywood school films and Hollywood films in general. Most mainstream movies contain a straightforward, narrowly defined underlying message that is reinforced by all of the movie's textual elements. The central convictions of *Dead Poets Society* (or of its creators) are definitely made manifest throughout the film, but at the same time the movie leaves some room—intentionally or not—for its true convictions to be questioned. Hence, the movie's apparently ambivalent positions could be interpreted as a slight step outside the Hollywood mainstream. After all, Schulman has spoken about the reluctance of many Hollywood types to accept his screenplay initially (Farley 1989). Yet, there is more that needs to be said about its divergent views. The question then becomes: What is (are) the film's position(s)?

It is admittedly difficult to make sense of these contradictory impulses. For one thing, *Dead Poets Society* demonstrates a belief in the value of pursuing individual creativity and freedom. As evidenced by the poetic verse quoted throughout the film, by the choices made by Keating's students, by the film's demonization of those characters (Mr. Nolan and Neil Perry's father) who most represent the oppressive side of conformity, and by its climax, the film celebrates the rejection of slavish adherence to norms. But the movie also shows reverence for tradition. The narrative seems to endorse nonconformity, but by concerning itself with an era before "drugs, the sexual revolution and controversies over so-called cultural literacy" (Holden 1989), it betrays its strong nostalgic impulses. The film's respect for tradition is also highlighted by its exclusive use of Wasp poets, students, and cultural influences.

If the film's climax is any indication of what was important to the filmmakers, then it would seem that the film sounds a cry against conformity. Yet, in the same climax, Keating seems to accept blame for that which is not fully his fault by not protesting his termination—certainly a tacit acceptance of tradition and rules. As Holden (1989) argues, "Whether or not it intends to, *Dead Poets Society* implies that the cost of going against tradition, even in positive, well-meaning ways, might not be worth the risk." To restate the point, there is some inconsistency in the film's responses to unorthodox behavior.

Ultimately, however, the movie promotes and symbolically applauds individual rebellion against "indentured servitude." *Dead Poets Society* wholeheartedly embraces the notions of seizing the day and making life extraordinary. But the movie's promotion of rebellion against conformity is narrowly focused on the individual and not on the collective good (Carton 1989; Holden 1989). The movie endorses the notion that "words and ideas can change the world," but avoids using its characters to effect any sort of larger social or institutional change. To be sure, *Dead Poets Society* is preoccupied only with *individual* freedom from oppressive forces.

Yet, despite the film's unwillingness to move beyond the level of the individual and call for broader institutional change, it remains critical of those institutions. As an example, some of the students and even Keating himself are punished for going against the grain; however, the punishers are made out to be the real wrongdoers. The orthodox behavior of the punishers is at best harsh: Perry's father forbids him from acting in the play, withdraws him from Welton, and decides Perry will attend a military academy then medical school, all driving Perry to kill himself; Mr. Nolan forcefully paddles Dalton for his prank, then later uses McCarthyite tactics to elicit accusations from the students against Keating; and Cameron, who reluctantly participates in the Dead Poets Society and is the first to blow the whistle on Keating, is portrayed as a traitor to the group, after which the others have no choice but to follow suit. And even though the students are punished in the end, the fact that they stand in support of

Keating at the very end illustrates their continued belief in his teachings. It seems these scenes are intended to criticize the oppressiveness of orthodoxy and to support the students' rebellion against it.

As for the film's other preoccupation, the filmmakers also appear to be singularly influenced by a homogeneous cultural tradition. As Canby (1989) asserts, Welton's "world is insular, that of the privileged white male who, if he is not already a scion of Old Money, will probably marry it." In addition, the school's curriculum is imbued with the spirit of "the classics": Keating's students also take courses in Latin, trigonometry, chemistry, and history. Yet the film's nostalgic impulses and fascination with tradition are not incompatible with its encouragement of nonconformity. In fact, the fact that the film is set in 1959 allows it to ignore using individual rebellion in the service of social change as it was used during the 1960s. This point will be crucial in the next section as the discussion focuses more specifically on the film's racial politics.

The Absence/Presence of the Other

The fictitious Welton Academy of *Dead Poets Society* is an all-male, all-white prep school in the rural Northeast. The academy serves as a meeting place for the sons of the well heeled and a pipeline to the Ivy League institutions. Its rustic setting offers all of nature's beauty, neo-Gothic architecture, and isolation from the mania of the rest of the world and particularly of the city.

The academy's physical isolation and cultural and socioeconomic homogeneity are important elements of the film's overall racial politics. The facts that there appear to be no nonwhite students or instructors at Welton and that the academy is distanced from the city and thus from the racial Other enable the filmmakers to avoid dealing directly with the issue of race and prevent the strength of their central message about individual rebellion from being diluted. The year of the story also seems to allow the filmmakers to excuse themselves from dealing with race, for as Carton (1989) states, "in 1989 mythology, 1959 is the year before racial struggle, sexual politics, poverty, war, drugs, decaying cities, and doubt began to appear in America" (64). And yet, Welton has no *female* students or instructors either, but this absence does not prevent the filmmakers from dealing with women. As a matter of fact, the absence of female students is rendered important in the film (albeit superficially so), but the absence of nonwhites is not.

However, there are two nonwhites—both African American—who appear, extremely briefly, in the movie. The first black person is seen on the local high school campus apparently after a football game. The camera focuses on the young woman that Overstreet desires, and in the background are a row of percussionists in the school's marching band. At the end of the row is a black male student drumming away. The second glimpse of a black face is in Welton's kitchen. The scene features Overstreet as he is returning from the local high school, where he went to court his love interest and sway her with a love poem. Having presumably missed a meal, he sneaks in through the kitchen and steals a piece of bread, innocently imploring the onlooker to keep quiet about it. As Overstreet turns the corner, a black man emerges and watches Overstreet skip down the corridor.

Thus, amid the virtual absence of any nonwhite people anywhere, the film includes two African Americans, and in stereotypical places—dealing with music and food. Perhaps the filmmakers meant to add to the film's credibility as a period piece by placing blacks where they actually might have been found, but they could have fully maintained the film's racial homogeneity and left nonwhites out of the film altogether. Instead, the filmmakers apparently chose to avoid both the messiness of race and the possible charges of ignoring racial difference by making nonwhite people part of the scenery.

Another way in which *Dead Poets Society* ignores racial difference and relies on cultural homogeneity is through its traditions. For one, Keating's students follow a classical curriculum. Their courses exemplify William Bennett's beliefs about what an ideal curriculum might look like (Bennett 1988). Moreover, all of Keating's and the students' "dead poets" are men of Anglo ancestry. In the academic curriculum, music, sports, and dress associated with Welton, the most immediate influences are British and Anglo-American. Welton's takes its traditions seriously; tradition is, in fact, one of the academy's principles. Outside of the fictional world of Welton Academy, the filmmakers of *Dead Poets Society* also seem to demonstrate a high regard for tradition, given their placing of the film in 1959, at a time before much of the behavior, values, and knowledge paraded in the film were fervently challenged in the sixties.

The above discussion points to the virtual absence of nonwestern traditions or even nonwhite bodies in the world of Welton Academy. Yet, there is an undercurrent of Otherness in the movie that must be explored. More specifically, when this undercurrent surfaces, it almost always plays a subversive role in the otherwise well-ordered world of the teen boys. Most of this subversion takes place in the meeting place of the Dead Poets Society. When the boys discover Keating's past membership in the society and inquire about it, he cautions them: "I doubt the present administration would look too favorably upon that." He then tells them that the society was dedicated to the celebration of poetry, and that the group met "at the old Indian cave" in the woods. Thus, the initial site of transgression is a dark, American Indian cave escaped to under the cover of night.

The passage that is read to open each of the society meetings—from Henry David Thoreau's *Walden*—signifies the cave's role as a place that the boys visit in order to truly "live." Carton (1989) points to the further symbolism of the passage in stating that Thoreau "repeatedly associates with the American Indian" the unadulterated life he seeks in the woods (65). The film also relies on other vague notions of "Indianness" in the music played in a dorm room scene in which one student plays a Native-looking wooden wind instrument and Dalton plays his hand-held drum. Finally, Dalton, the most irreverent youth of the group, adopts the name "Nuwanda" and demands that the rest of the boys refer to him as such from that point on.[3] The name is of unspecified origin, but it certainly is not a typical name for a white male in 1959, and it is undoubtedly intended to suggest something non-European or even primitive. Giroux (1993) argues that Dalton's act is "a striking display of colonialism" in which he "takes on the identity of the Other as the province of the primitive, exotic, and romanticized warrior" (45). Later on, as the boys are dressing to attend Perry's play, Dalton emerges from a bathroom stall with what appears to be a bright red lightening bolt painted across his chest, which he claims is "an Indian warrior symbol of virility." Based on these examples, it is evident that the use of supposed American Indian symbols and cultural artifacts is meant to correspond with an unleashing of one's inhibitions and true nature.

The same appears to be true of the influence of blackness in the movie. The cave is pregnant with symbolism around the idea of darkness; for instance, the boys first sneak to the dark cave, in the dark of night, in dark, hooded, duffel coats. In one scene, Meeks recites a portion of Vachel Lindsay's "The Congo (A Study of the Negro Race)." Importantly, he recites from the first section of the poem, subtitled "Their Basic Savagery." The boys start clapping to his rhythm, then start reciting in unison the following lines:

> Then I saw the Congo
> Creeping through the black
> Cutting through the forest like a golden
> track

From this point, they start their "act": one of them starts pounding on a garbage pail; they rise as a few of them make animal-like sounds; others start beating sticks together, Pitts plays a Jew's harp, and they begin marching in a circle. The scene climaxes with them processing out of the cave and into the woods in their new, primal state, complete with African undertones; perhaps this is the film's interpretation of the "basic savagery" and "irrepressible high spirits" of the "Negro."[4] During a later meeting in the cave, Dalton plays jazz passages on a saxophone between citing verses of original poetry. Though now seen as America's "classical" music, jazz originated, of course, among black Americans and was once seen as unrestrained and noted for its lack of observance of the classical tradition. These two incidents, along with the brief glimpse of the African American high school drummer, stereotypically connect blackness to musicality and to unrestrained, expressive behavior.

Music, the cave, and non-Western cultural influences, then, are in some way involved in the boys' most important acts of rebellion against Welton's traditions. First, the environment of the cave allows the boys to express their most primal, natural urges. This is highlighted in the comment made by Headmaster Nolan that the boys used the cave to engage in "recklessness and self-indulgent behavior." Second, their new uninhibitedness is expressed in part through music and dance borrowed from other racial groups. Third, the appropriation of non-Western cultures symbolizes their break from orthodoxy. These three elements are powerful forces of opposition to the sterility and order of the Welton environment.

Consuming the Other and Spitting Out the Seeds (of Revolution)

There remains, however, the film's celebration of an unadulterated Western tradition safely isolated from all of those non-Western influences. It is easier to understand the film's simultaneous fascination with tradition and rejection of conformity when the film's messages are examined through the lens of race. Reeves and Campbell (1994) describe a "general moral paranoia of the New Right" that undoubtedly fuels (at least in part) the opinions held by those like Allan Bloom and William Bennett:

> In the moral universe of the New Right, the college campus is an out-of-control space where generation after generation of young people challenge authority, reject traditionalism and "family values," embrace hedonism, and court radicalism. (154)

I submit that these four acts are committed by the youths in *Dead Poets Society*. What is more notable about this is that these acts are not totally suppressed in the film's narrative. On the contrary, these same elements, which inspire fear in many right-wingers and are condemned in the multiracial school films discussed in chapters 3 and 5, are actually celebrated in *Dead Poets Society*.

The presence of non-Western influences amid the absence of actual non-Western or nonwhite people seems to allow the filmmakers to focus on white people while avoiding the messiness of race, and allows them to exploit those unseen others in the service of liberating those whites. Carton (1989) makes a similar point when he argues the following:

> For as [this neglect] disqualifies certain groups from full human subjectivity, it also makes them available to be possessed as commodities—or more specifically, it makes available for commodity consumption those of their stereotypical attributes that white males might associate with their own release into nature. (65)

The movie permits—or actually encourages through the character of Keating—the teen boys to flirt with unorthodox behavior and exploration.

Along the same lines, Farber and Holm (1994) argue that in *Dead Poets Society*, "where the students already are high achievers on the track of success, we witness them as they gain a new definition of success at the top of the social-class hierarchy, *one that frees them to really live and achieve full standing as human beings*" (170; emphasis added). This is unlike what happens in the school films that deal with nonwhite youths. In *The Principal, Lean On Me*, and later in *Renaissance Man* and *Dangerous Minds*, the nonwhite students are commanded to follow rules, orders, and regulations or are urged to break the rules *only* under the supervision of the (usually white) authority figure. Curiously, though, the filmmakers stopped short of letting the students in the Dead Poets Society "make their lives extraordinary" by fighting larger forms of institutional oppression. Although they found the idea of teen rebellion intriguing, they were perhaps afraid of endorsing rebellion too fully lest the audience (especially the right-wing segment) perceive Keating as planting the seeds for the growth of civil rights demonstrations, university sit-ins, antiwar protests, and sexual revolution.

Restricting the appearance of nonwhites to only the briefest of glimpses also permits the filmmakers to treat the white boys as victims of oppression without having to justify not dealing with the oppression of others.[5] Some of the film's language seems to support this point. The first instance occurs during a conversation between Keating and Perry about his dilemma over appearing in the play. Perry is uncomfortable about defying his father's wishes by remaining in the play, so he seeks Keating's advice. "You are not an indentured servant," Keating tells him, and suggests that Perry express his desires to his father who will eventually understand. In the second instance, which happens close to the end of the film, Cameron is challenged and physically threatened by the other boys who accuse him of incriminating Keating before the headmaster. Cameron defends himself by exclaiming, "We're the victims." In these scenes, the white youths employ the language of victimhood for which the nonwhite students in *Lean on Me* and *Dangerous Minds* are censured for using. The filmmakers seem to be saying, "Look, White men can be oppressed, too—and here's how." This is a crucial step in the full recuperation of white men from the blows suffered during the sixties and seventies—which, again, are also too messy for the movie to handle.

Let us return to the contradictory impulses in *Dead Poets Society*: the upholding of (Western) tradition and the celebration of rejection of traditionalism. In light of the discussion in the preceding sections, it appears that these two factors do and can exist in reasonable harmony in the same narrative. The movie strikes the following compromise: that one can venture outside the canon of traditional thoughts and deeds as long as one is first fully grounded in Western tradition unmerged with other cultural influences. Thus, the movie holds on dearly to its traditions while allowing its characters some amount of experimentation outside of them.

The denizens of Welton Academy—white, male, class-privileged—are told in the narrative of *Dead Poets Society* that they can be freethinkers and nonconformists. And why not? After Welton, the Ivy League, and the professional school of their choice, the world will be theirs. This, of course, is not the case for their "less civilized" peers. The largely working-class, nonwhite, troubled young men and women of the other school films, who are constantly tempted by the snares and dangers of the city, cannot be allowed such freedoms. According to this view, it is the lack of restraint and order that is the source of most of their problems in the first place. Some might muse that the best way to civilize them is to bring them back under the influence of Western knowledge and culture. This is an idea that the school films in the next chapter explore in detail.

5

Bringing the West to the Rest: The Transmission of Civilizing Knowledge

On the surface, the first election of William Clinton appears to have signaled a major political and cultural shift in America, or at least a rejection of the Reagan–Bush legacy. Clinton's platform certainly featured rhetorical repudiations of Reagan–Bush policies; Procter and Ritter (1996) argue that his inaugural address, for example, employed "regenerative rhetoric" (in the form of the political jeremiad) that included as one of its three themes an undermining of the community constructed by the Reagan and Bush administrations. Even more, his early calls for health care reform and nondiscrimination against gays and lesbians in the armed forces gave the electorate good reasons to expect his political wind to blow in a different direction. Yet Clinton refused to make a clean break with the conservative politics of the preceding twelve years. Once in office, his alternately liberal, centrist, and conservative decisions earned him much fascination and criticism in the press; moreover, he soon became stigmatized for his "flip-flops" and wishy-washy behavior.[1] Based on these observations, the descriptor "vacillating" might have applied to Clinton's leadership equally as well as the centrist description.

Clinton had already established himself as an intellectual and political leader of the Democratic Leadership Council (DLC). From the position of chair of the DLC, he advocated for a new direction for the Democratic Party that he described as both liberal and conservative, rejecting the restrictive alternatives represented by the old Republican–Democrat categories (Clinton 1991). However, Clinton's attempts to become a "New Democrat" ultimately placed him more in step with Reagan–Bush racial politics than might have been expected by the political pundits and the American electorate in 1992. In fact, given the fact that many Americans perceived Clinton as an "ideological shape-changer who becomes what each beholder wishes to see" (Kelly 1992, 36), few voters knew what to expect. A similar analysis of his rhetoric describes it as constantly mutable, displaying an ability "to shift political positions, to change primary colors, and to steal the Republicans' fire even as he gives them a bipartisan Trojan horse" (Murphy 2002, 233).

Amidst Clinton's rhetorical mutability was one constant factor: his verbalized repudiation of the Reagan–Bush legacy. Yet, despite Clinton's promises to the contrary, there are several "unsettling continuities between the rhetorics of Reaganism and Clintonism," resulting in some cases in a repackaging of Reagan backlash (Reeves and Campbell 1994, 251). Reeves and Campbell (1994) point to one of these continuities in the following statement:

> But, ironically, in the midst of his condemnation and demonization of the Reagan order, he also reaffirmed a central principle of Reaganomics: Clinton's promise to eliminate 100,000 bureaucrats from the federal payroll and put 100,000 more police on the streets is not only a ringing endorsement of the police state, but it is an encoded pledge to continue the Reagan-Bush administrations' discriminatory program of deregulating entrepreneurial interests while exercising even more discipline on reserve and criminalized labor markets. It was a pledge that should have warmed the cockles of Daryl Gates's heart. (252)

Ultimately, the crime bill that passed contained Reaganite elements such as authorization of the death penalty for an expanded list of crimes, a "three-strikes" sentencing policy, and a provision for trying juveniles as adults in federal courts; and it contained a more liberal element in the form of an assault weapons ban (Quirk and Hinchliffe 1996). Although the end result may appear to be a centrist's compromise, the link to Reagan era politics is undeniable.

President Clinton's 1992 campaign promise to "end welfare as we know it" also vividly recalled the rhetoric of Reaganism.[2] Admittedly, his solution to the welfare "problem" was different: he called for education, skills training, and child care instead of a simple elimination of benefits or drastic restriction of eligibility. But he failed to offer a proposal in time for it to be considered before the Republican takeover of Congress in 1994 and lost the upper hand as a result. Afterwards, he faced political pressure to accept one of the Republican welfare bills. Fearing a drop in public opinion polls in the lead-up to his reelection, Clinton moved beyond his campaign rhetoric and upheld the Reagan–Bush legacy by signing the 1996 welfare reform bill not long before that year's election, even though there was no reliable indication that winning the election depended on his signature (Edelman 1997; Quirk and Cunion 2000).[3]

With regard to the economic realm, we can see again Clinton's laying out of, or seeming to lay out a "new," centrist path. His "vision of change for America" was sold as a corrective to the economic failures under Reagan and Bush. This vision, though full of details, involved two main aims: stimulating the economy and reducing the deficit. The deficit reduction strategy took the form of the Omnibus Budget Reconciliation Act, which passed in 1993, though barely so. On the other hand, the stimulus package was abandoned shortly after a Republican filibuster blocked it in the Senate (Meeropol 1998). What looked like a partial victory for Clinton actually masked, in Meeropol's words, "the important fact that the Reagan Revolution had succeeded in shackling even a reform-minded Democratic president supposedly working with a like-minded Democratic majority in Congress" (236). After the loss of the Democratic majority in 1994, Republicans assumed the lead role on deficit reduction (Quirk and Cunion 2000). By using the leverage of their new majority and the language of their "contract with America," the Republicans were able to force Clinton to agree to a deadline for a balanced budget. With his submission of his budget and his signing of the welfare reform bill in 1996, Clinton "signaled surrender" to Reaganomics, or more specifically to the Volcker–Reagan policy revolution (Meeropol 1998, 6).[4]

The Racial Politics of the "First Black President"

Racially speaking, it would be most unfair to claim that Clinton simply maintained the politics of the Reagan era. This was, after all, a man who appeared on "The Arsenio Hall Show" and played blues music on a saxophone, and who asked Maya Angelou to write and recite the poem for his first inauguration. These are just a couple of nonpolitical examples (or perhaps eminently political ones), but they underscore a larger point: the cultural resonance of involving and connecting with those two African American figures was significant and certainly served to distinguish him from his two predecessors. There are, of course, numerous ways of accounting for the differences in his approach to racial issues from that of Reagan or of Bush. One explanation suggests that because he was the first president to have come of age in the midst of (and after) the civil rights movement, his thinking about race was shaped within a cultural context that neither Reagan nor Bush had experienced.[5] In addition, Clinton had been exposed to and even developed meaningful relationships with professional blacks and other nonwhites during his academic, legal, and political careers; these were people to whom he could and did look for advice and friendship—and from among whom he could and did find workers for his administration.

And Clinton's different experience with, or understanding of, or approach to issues of race was reflected in many of the political appointments and decisions he made. Martha Farnsworth Riche observed that, by 1994, Clinton had made good on his campaign promise of having an administration that "looks like America." As of December 1993, with more than 75 percent of his positions filled, the President had created an administration that closely mirrored the racial and ethnic makeup of the United States. Whites, blacks, Latinos, Asian Americans, and American Indians were almost proportionately represented—with blacks being slightly more so and Latinos being less so. Riche notes how remarkable an accomplishment that was, "especially since the demographics of the pool of eligible people . . . is considerably less diverse than the general population" (1994). As she appropriately points out, the status of those appointees was a whole other important issue; furthermore, the "rainbow of hues" was not "matched by a rainbow of views" (Horne 1994, 187). But the simple diversity "bean count" signaled that Clinton had a different agenda regarding race.

His symbolic moves also demonstrated a departure from the Reagan–Bush legacy. Shull (1999) contends that Clinton changed the civil rights focus from what it had become under his Republican predecessors and that he was also the most attentive president ever to civil rights (though his statements were generally untargeted). The Clinton administration's public opposition to the unprecedented initiative to ban affirmative action in California (Proposition 209 of 1996)—both during his 1996 campaign and by joining the court challenge of the proposition—is only one example of its divergence from the anti-affirmative action sentiment and maneuvers of the right wing (Yeager 1996). President Clinton's Initiative on Race, launched in 1997, was also a push for a public discussion of racial issues that had no precedent under Reagan or Bush. Even though the initiative was beset by disagreements over its purpose (Franklin 1998) and though legislation was slow to emerge from the initiative (Shull 1999), it was a step that set him apart from anything the other two leaders had any interest in taking.

Still, this was the same Clinton who during his first campaign had deliberately marginalized Jesse Jackson (in part by using rapper Sister Souljah as a political pawn) and knowingly played golf at a racially segregated country club in Arkansas. Later, during his second term, he declined to offer a formal apology for slavery and opposed any reparations even for descendents of slaves (Shull 1999). Also, prior to his eventual opposition to Proposition 209, Clinton's chief of staff reportedly was "quietly asking friends on Capitol Hill whether the president should simply endorse" the initiative (Roberts et al. 1995, 35). Clinton is notorious among citizens of various political persuasions for his avowed desire to "mend, not end" affirmative action, a statement at once meaning anything and nothing. The statement can easily be seen as merely an attempt to appease those on the Left without losing the support of social conservatives. It typifies Shull's (1999) conclusion that Clinton was the most rhetorical yet least supportive (i.e., least liberal) Democratic president in his public statements on civil rights. While not a mere repeat of the two previous administrations, Clinton's reign did display the impact of the racial conservatism of the Reagan–Bush legacy and yielded a minimal civil rights agenda (Walton 1997). Walton and Smith even claim that the regulatory reform under Clinton had no civil rights thrust and that "*[i]n both the long and the short run, the Clinton regulatory reform may impact the African American community more negatively than the Reagan and Bush ones*" (2000, 254; emphasis in original).

Nonetheless, Clinton's actions did not cost him (at least in some ways) as much one might imagine among nonwhites. For instance, he increased his proportion of the vote among Asian Americans and Hispanics between his 1992 and 1996 campaigns and maintained his high level of loyalty among African Americans (Walton and Smith 2000; Judis and Teixeira 2002).[6] Interestingly, blacks were particularly supportive of Clinton during the Monica Lewinsky scandal and subsequent impeachment process; public opinion polls indicated that over 90 percent of blacks opposed impeachment (Walton and Smith 2000). It was in the midst of this support that African American comedian Chris Rock called Clinton the "first black president." To be sure, the comedian captured the affinity that many blacks seem to have had for Clinton. At the same time, Rock's commentary was satiric: for him, Clinton was "black" because "they question everything he does"—in other words, he was the subject of the same undue scrutiny under which many blacks see themselves as constantly living.[7] Despite the fact that several of Clinton's gains had been won at the expense of black advancement, he maintained generally high approval ratings among the group over the course of his entire administration.

In sum, Clinton's racial agenda displayed the promise of his New Democrat approach, as well as relied on the race-baiting used in the past. He set out to reverse some of the changes that had been wrought under Reagan and Bush, but his more moderate policies seem to have been more successful in the long run. This success was a mixed blessing, however, for it frustrated and alienated some racial minorities who had hoped for a more liberal legacy from a man who demonstrated such apparent ease with racial matters. The Congressional Black Caucus, for instance, experienced some feelings of betrayal due to its belief that Clinton was reacting too much to Republicans, as evidenced by, for example, the versions of the welfare reform and crime bills he endorsed (Quirk and Hinchliffe 1996; Walters 2003). Sadly, O'Reilly concludes, "arguably the least prejudiced" of the presidents saw fit "to include a racial calculus in politics and policy" (1995, 420).[8]

Clinton on Education

In October of 1989, President George Bush participated in a meeting with the nation's governors—which included then-Governor Clinton—for a summit on education, only the third domestic summit in the nation's history (Cavazos 2002). Out of this summit emerged a set of performance goals that were folded into what would come to be known as Bush's "America 2000" plan. Instead of discarding this plan when he entered office, Clinton and his administration repackaged the agenda and sold it as the "Goals 2000: Educate America Act." The act was approved by Congress in 1994 (Stallings 2002). It could be argued that Clinton merely adopted the Bush agenda that preceded him, but this would be too simple a conclusion to draw in this case. True, Clinton and his Secretary of Education, Richard Riley, appear to have espoused the questionable contentions made in *A Nation at Risk* that American education was increasingly mediocre and at a point of crisis (Clinton 1996; Riley 2002)[9], beliefs that put them in the same school of thought as Reagan. However, Clinton, as a part of the National Governors' Association, had helped to hammer out these goals during the 1989 summit; thus, he had some stake in their success and apparently believed that they were viable when he was a governor and, later, as president. Moreover, the act built on and aligned itself with reforms that were already taking place in states and communities (Smith et al. 1997; Riley 2002), which made for politically and programmatically smart legislation and not just warmed-over Bush administration policy.

Clinton's liberalism was more evident in the educational arena than in other arenas as his administration moved beyond the Bush agenda by renewing and increasing funding for critical programs, federal grants, and loans, and by introducing higher education tax credits (Stallings 2002). Another example is the criticism from the 1992 Clinton campaign via the Democratic Party of Reagan's and then Bush's attempts to "bankrupt the public school system" by offering vouchers for private schools (Greider 1992, 25; see also Roy 1997). The administration also made significant legislative accomplishments early on in Clinton's stay in the White House, legislation that featured the flexibility to support local and state efforts rather than a demand for compliance with rigid regulations. Overall, the Clinton educational agenda continued the federal government's traditional emphasis on assisting and protecting those most in need in society (Smith et al. 1997).

The analysis by Smith and others (1997) of the educational policy approaches of Clinton's first term provides evidence for the argument that his Department of Education consistently combined a wide range of public campaign strategies with a successful legislative agenda to support a clear set of reform goals. While not an entirely new approach, the degree to which his administration engaged in it departed significantly from prior practice. Even more pertinent here is their conclusion regarding this practice: "Rather than being inconsistent with traditional Democratic strategies and goals to provide educational opportunities for all children, we believe these other approaches have enabled the DOE to focus on the kinds of issues—and become involved in ways—that support those goals" (Smith et al. 1997, 223–224). In short, Clinton's education agenda, more than most of his major policy agendas, gave priority to and fought for ideas that were typically associated with Democrats, such as strong support for public schooling, federal grants, and governmental safety nets for the most needy. At the same time, the agenda embraced plans like national education standards that up to that point had been more favored by Republicans. The administration's actions regarding education are in keeping with the "new" approach Clinton advocated to great effect back in 1991, an approach that "rejects the Republicans' attacks and the Democrats' previous unwillingness to consider new alternatives" (Clinton 1991). In this realm, Clinton appears to have come closest to honoring his "Third Way" or New Democratic approach to politics.[10]

And what of this Third Way approach? Quirk and Cunion's examination of Clinton's New Democrat politics is insightful on this matter: "A centrist approach can produce highly effective campaign rhetoric. At least in the current state of party politics, however, it is more problematic as a strategy for governing" (2000, 201). This analysis echoes Reich's (1999) warnings of how promising—and perilous—Third Way politics can be. In this way, perhaps Clinton was doomed to fail in his effort to lead a centrist administration. It did not help him to have been faced by an oppositional, Republican-dominated Congress after the 1994 mid-term elections, a position that placed him between the Scylla of gridlock and the Charybdis of acquiescence. Clinton, ever the "brilliant political tactician," survived this tricky day-to-day business, of course, but he did so without an overarching strategy (Rockman 2000, 292). Indeed, it is questionable whether he even wanted something more significant than mere political survival (Rockman 2000). Far from turning "the Third Way into a moral crusade" (Reich 1999, 51) or new political path, Clinton teetered precariously (if adroitly) between the old categories, but capitulating all too often to the continued influence of Reagan's legacy and refusing to sever his ties with it.

"Cinema of Recuperation" and the "Recovery" of White Supremacy

As expected, the Reagan–Bush influence is also perceptible in the Clinton-era cultural arena, specifically in more recent Hollywood movies. The reactionary racial politics that gained momentum during the Reagan era were still at play on the silver screen in the 1990s. But, as times change, so do the creative expressions that emerge out of those times. In many ways, the filmic texts that have been produced since the first election of Bill Clinton resonate with some of the liberal positions for which he stands. In the newer political and cultural climate, however, Reaganite ideologies parade rather smoothly along the roads previously paved by the racially coded imagery disseminated during the 1980s. This is possible as long as their conservative messages are disguised behind a more liberal mask.

Some scholars have identified a phenomenon of racial "backlash" in Hollywood films of the past 20 years, collectively referring to these films as the "cinema of recuperation" (Wood 1986; Guerrero 1993).[11] Guerrero (1993) identifies the political underpinnings of this body of films and the particular implications with regard to African Americans in the following statement:

> In the beginning of the 1980s and under the political impulse of Reaganism, blacks on the screen, in front of and behind the camera, found themselves confronted with the "recuperation" of many of the subordinations and inequalities they had struggled so hard to eradicate during the years of the civil rights movement and the emergence of Black Power consciousness that followed it. (113)

If we locate the ascent of this backlash in the period from the late seventies to the beginning of the eighties—as do Guerrero,

Wood, and others—then we should recognize that it continued through that decade, and reached its highest point in the 1990s during the generally more progressive environment of the Clinton era. Instead of being suffocated in this environment, the backlash animal has gained new life in reaction to it.

At the beginning of his text, *The Recovery of Race in America*, Aaron Gresson (1995) discusses the human tendency to invest in various myths or narratives that aid us in making meaning of our everyday lives. When these myths are unsettled and emptied of their ability to explain the world, individuals "recover" from this loss of myth by creating or buying into new myths, or explanations, to replace the old ones. In reference to this process, Gresson (1995) says the following: "I call this task of individual and collective persuasion 'the recovery project,' by which I mean that people must recover ways of being related and connected to something and someone larger than 'I' and 'me' " (Gresson 1995, 3).

Since we are all susceptible to individual loss, any one of us may attempt to latch on to new narratives in order to deal better with what is lost and to personally extricate ourselves from this bankruptcy of myth. Yet the process of recovery, as a human phenomenon, is obviously not exempt from the exercise of power and authority between individuals and groups. Indeed, "[p]ower plays a central role in recovery rhetorics, mainly evident in the right to name, to define, and to self-validate choices affecting others as well as oneself" (Gresson 1995, 5). With power added to the formula, recovery can be seen as "the reconnection to, or regaining of, a prior and privileged position or relationship" (4). Several groups have in recent years set off on their respective journeys of recovery. The rhetoric of numerous men in reaction to the rise of feminisms—which branded feminist women as "man-haters," "lesbians in disguise," and "femiNazis"—can be thought of as their attempt to reassert male supremacy and protect patriarchal domination. At issue in this study, however, is the recovery of racial positionalities.

To state the obvious, films center on narratives and transport myths; therefore, the process of recovery is clearly applicable to Hollywood texts. The school films under examination in this chapter might be seen as collectively contributing to what Gresson calls the "white recovery project," which is an attempt by whites to recover a real or perceived loss of power (1995, ix). Like Wood (1986), Gresson locates the genesis of this period of recovery in the 1970s. Unlike the eighties and early nineties when its interests were supported by the White House, the right wing in the 1990s had to adapt its tactics accordingly. More to the point, the white recovery project has found the Hollywood film industry to be a Trojan horse in its desperate attempt to reassert its racial dominance.

Although the earliest recuperation films began to appear at the beginning of the Reagan era, the nature of the nineties school films differs from that of their older filmic kin. I consider the school films to be discussed in this chapter to be the third step in the sophisticated process of recuperation. In these earlier films, I have noted both the emphasis on the dangerous threat of the nonwhite body and the urban space as well as the emphasis on the civility of white students and the serenity of suburban life. Thus, the first step involved the demonization of nonwhites, whereas the second involved defining the image of whites in diametric opposition to the demonized portrait of the Other. Now, in the most recent installment of school movies in the more Left-leaning Clinton era, whites are shown as saviors who not only tame the threatening Others, but restore order to their habitat and civilize them.

Lola Young (1996) writes about what she sees as a recurring feature of movies made by white people: "a constant refusal to relate intimately to black people's knowledge and experiences, despite protestations to the contrary" (26). But the nineties school films attempt to contradict Young's observation. These films might even be accused of flaunting their liberal credentials, for their narratives pretend to sympathize with and understand the plights of their multiracial subjects; however, this appearance disguises the fact that they often prescribe some pretty conservative remedies to their problems. Three of these school movies wear the liberal disguise well: *Renaissance Man*, *Dangerous Minds*, and *Music of the*

Heart. Contrary to what they appear to be, the movies typify the most recent attempt by conservative interests to recuperate from the blows to the racial positionalities once held by them and now perceived to be lost.

A "Classical" Tale of Uplift

Imagine the following scenario: A hard-working, gifted professional loses his job and runs into difficulty finding similar work. As a result, he accepts a teaching position with a class of supposed slow-learners quite different from himself. He is initially unable to connect with them, much less teach them, until something unexpected occurs and sparks the students' interest. From that point on, the students exhibit a strong and sentimental attachment to their teacher, demonstrate their thirst for knowledge and capacity for learning, and persuade the teacher to embrace his newly discovered profession.

The above sketch may bring to mind the film *To Sir, with Love* (1967),[12] but it is an equally appropriate synopsis of the more recent *Renaissance Man*. Though its basic story line greatly resembles the older film, *Renaissance Man* diverges both aesthetically and ideologically from its predecessor. In the case of the latter movie—which is largely a comedy—the lost job is in advertising, the campus is located on an army base, and the students are recruits whose commanding officer has determined that they need "basic comprehension" skills in order to become capable soldiers. Also important is the fact that this time, instead of a black male instructor being thrust into a class full of white working-class youths in which race was a contentious issue between students and teacher, the instructor is a white man for whom race is not outwardly an issue, struggling to deal with a smaller group of geographically and racially diverse students.

What's in a Name?

Given the subject matter of *Renaissance Man*, the above line from Shakespeare seems wholly appropriate as an inquiry about the significance of film's title. It seems that the film provides at least three possible motivations for its title. The first and most probable motivation is that the filmmakers saw the teacher and his students as "Renaissance men" (or a Renaissance woman in the case of Miranda) as a result of their enriching military experience. In order to achieve this status, the students (and the teacher) had to develop their talents in more than one arena, following the model of "universal men" like Leonardo da Vinci and Leon Battista Alberti to whom the term often refers. These men possessed a wide range of interests and enormous talent, and the suggestion in the film is that the students should be proud of their accomplishments both as students and as soldiers. Second, it is likely that the film's title is a play on the fact that the teacher's former place of employment is located in a downtown Detroit complex called the Renaissance Center. A third possible explanation accounts for the fact that the texts that changed their lives were penned by Late Renaissance author William Shakespeare.

The plot of *Renaissance Man* is reportedly based on the experiences of the man responsible for the script, Jim Burnstein. He apparently had some success teaching Shakespeare to enlisted personnel in the National Guard, and this, his first screenplay, grew out of his experiences (Lowry 1994; Maslin 1994; Wilmington 1994). It is unclear how much of the film is directly related to Burnstein's experiences, but he and the other filmmakers obviously decided that his story would be entertaining to a larger audience. In light of this decision, let us examine how the filmmakers draw upon this celebrated era of Western civilization in order to rehabilitate a multicultural squad of urban and rural soldiers with "special needs."

The main character of *Renaissance Man* is Bill Rago, an award-winning adman whose hostile behavior and substandard work cause him to be fired. Out of desperation and out of a need to support himself and his teenaged daughter, Rago ventures to the unemployment office. The film's racial politics surface immediately. Along the way he passes a white vagrant while crossing the street. He also passes by five African American men and one African American woman singing for money on the corner outside the entrance to the office. Evidently these people are unemployed, but for whatever reason are not *inside* the office seeking employment. Perhaps their little ditty explains why: "I'm flat broke but I don't care." When Rago reaches the help desk, he encounters a black female unemployment officer with whom he establishes a comically antagonistic relationship. He departs with her telling him that the office will contact him in a couple of weeks.

On Rago's next visit to the unemployment office, he again passes the singing panhandlers in the same place, yet this time he extends a friendly hand and greets them (maybe this is the filmmakers' way of conveying Rago's humility and connection to the common folk). By this point his relationship with the black female worker has improved considerably. She informs him that a job has been found for him—as a teacher. After he declares his unsuitability for the job, she tells him, "You have a master's degree [from Princeton]. That means you can teach."

Rago learns at the unemployment office that he will be expected to live and work for six weeks at an army training camp, but does not discover until arriving at the base that he will be teaching "basic comprehension." From the moment he drives onto the base of Fort McClane and has trouble finding his way to his assigned post, the audience learns that military life is quite alien to the securely middle-class civilian lifestyle to which Rago is accustomed. The following exchange with between him and his supervisor, Captain Murdoch, is indicative of Rago's feelings toward the military: After reading an article on the subject, Rago sarcastically says to Murdoch, "That Gulf War thing worked out for us, huh?" Murdoch responds contentedly, "Sure did." He has the opportunity to absorb the military base's inspirational mantras—such as "Valor . . . Stamina . . . Integrity," "Pain is temporary: pride is forever," and the ubiquitous Fort McClane motto, "Victory starts here"—and finds himself in strange territory. He later expresses his antimilitary beliefs in offhand comments like describing the army as an "institute of imbeciles" and using the term "military intelligence" in his class as an example of an oxymoron. Rago's reactions to his Spartan living arrangements, to a 4:30 AM reveille, and to the regimented lifestyle of the recruits further demonstrate his opposition to whole idea of military training.

Dramatis Personae

Rago's students are a troop of rookies disparagingly referred to as the "Double Ds"—"dumb as dogshit"—for their lack of the requisite "smarts" supposedly needed to comprehend the orders of basic training. The students have been placed in the remedial class at the behest of commanding officer Colonel James, who believes that any person who voluntarily enlisted in the army should be allowed to stay and serve their country. When the class begins, one of the first assignments is for each recruit to write about why he or she decided to join the army. Afterwards, as each of the students reads his or her autobiographical text aloud, he finds that their life experiences are as foreign to him as the military environment in which he has landed. Amid the dissing between the students as each attempts to tell his or her own story, Rago learns as much from what they say about themselves as from the other students' commentary about them.

Here are brief descriptions of the eight recruits as gleaned from their reports: (1) Miranda Myers is a black female, most recently from Detroit, who has lived in several cities with her single mother. One day her mother abandoned her in Atlanta to follow a man. She claims that the "Be all you can be" army promotional poster inspired her to join and find out what she could become. (2) Donald ("Donnie") Benitez is a white male from New York City whose sister was accidentally shot to death. After that incident, he "decided that being a part of a gang didn't seem like such a good idea, so I joined this gang"—the U.S. Army. (3) Tommy Lee Haywood is a white male from Willacoochee, Georgia. His decision to join the army was based on the fact that he wanted to explore the world outside the one provided by the trailer park in which he lived. (4) Brian Davis, Jr. is a white male from Grand Forks, North Dakota with powerful emotional ties to a father he only knows through his mother's memories. He joined the army as a tribute to his father who died in combat in Vietnam. (5) Roosevelt Hobbs claims that he "wanted to learn a new trade." He is an African American man from Detroit who exhibits an air of coolness and introspection. (6) Black male Jackson Leroy is a former high school football star from Detroit. "I guess the real reason I joined," he says, "was to get a better education so I could get a better job." (7) Melvin ("Mel") Melvin, a white male from Charlotte, North Carolina, has a chronic sleeping problem; in fact, he sleeps through the assignment. It is later revealed that he was a victim of abuse by his stepfather. (8) The last is an African American male, Jamaal Montgomery, a perpetual wisecracker also from Detroit. He joined the army for the consistency and order that was lacking in his home.

At this point in the narrative almost all of the characters have been introduced, including Rago's daughter, who yearns for her divorced father's support for her aspirations. The only character yet to be seen is Colonel James, the white male head of the base who is as pleasant as Captain Murdoch (both being atypical portrayals of military officers on film). Now that we know who the players are, let us deal with the specifics of the story's unfolding. This will allow for a closer analysis of the movie's messages about the military, about education, and about race.

The Rago Curriculum

Once Rago has received Captain Murdoch's vague directive about what his students are to accomplish and has acquainted himself with them, he is more aware of his role in the training process of the eight recruits. Nonetheless, he has no idea what to teach to the students or how to proceed. Having a student who has a chronic sleeping problem and another who is a perpetual jokester presents a formidable obstacle to an instructor who has never taught before or likely spent much time around such characters. Rago almost gives up on trying to teach the "hyenas" until the ever-curious recruits interrogate him about his own reading material—Shakespeare's *Hamlet*. He claims that the play is one of his favorite works of literature and tells them that it "pretty much beats the heck out of any book ever written." Even though he has to explain to them what a play is ("Like TV . . . without the box"), once he describes the play as being about sex, incest, and murder, the students beg Rago to use *Hamlet* as the classroom text. He initially hesitates, thinking that the play may be too complicated for them; short of any better ideas, however, he acquiesces to their request.

Some reviews of *Renaissance Man* remarked on its commonalities with its Disney cousin, *Dead Poets Society*. While there are certain elements that the two films share, there is a feature of *Renaissance Man* that distinguishes it from *Dead Poets Society* and from most of the Hollywood school films to date. In the majority of the school films, the educational intervention is successful because it is unorthodox, because it *appears* to break with tradition in order to fill a void. In *Dead Poets Society*, Keating has his charges commit the controversial act of ripping the introductory pages from their literature text; in *Lean on Me*, Joe Clark uses a bat and bullhorn to establish order; even in *To Sir, with Love*, Mark Thackeray steps out on a limb and treats his students like adults. But *Renaissance Man* tells the story of a teacher who introduces a canonical author and his texts to his students in order to liberate their minds. It may be unorthodox and appear nontraditional in the context of a military institution to teach Shakespeare, but, in the context of American education, just the opposite is true. In this way, the narrative makes a link to the past, suggesting a Reagan-like "return to tradition" as the path to success.

The manner in which *Hamlet* is selected as a text for the class is worthy of a further comment in order to highlight the film's liberal veneer covering its traditional foundation. Rago appears to be progressive by allowing the students to choose their own reading material; instead, they choose to study Shakespeare. The film therefore avoids imposing the canonical author on the unsuspecting students. We know, of course, that it is the filmmakers who made the choice for the characters and for us.

Rago's excitement about the play shines through as he starts to read aloud from the text to the class, but the students are slow to catch on: expectedly, they encounter problems with the language. Their incomprehension leads Rago to attempt to explain the literary devices being used in Shakespeare's famous tragedy. In other words, Rago actually begins to teach—and successfully, judging by the way the recruits are able to apply his lessons to their lives. Shakespeare's language appears to exert the magical force over the students that it has been credited with doing for centuries.

Based on the remarkably low expectations that were held of the Double Ds at the movie's beginning, Rago, Capt. Murdoch, Sgt. Cass, and even the audience should be surprised at the amazing progress they make in the movie. Halfway through the film, the recruits decide to add their own flavor to the venerable text and encapsulate the story of *Hamlet* in snippets designed to match the cadence of their drill marching. Led by the African American recruits, the class also creates a rap version of *Hamlet* to perform for Rago. Later in the film, Rago arranges to have them attend a professional performance of *Henry V* (which the students also studied) in nearby Canada. They are all captivated by the performance, and even Melvin stays awake and even attentive throughout the entire show. Finally, in a partially climactic scene, Benitez delivers a speech from *Henry V* to Cass and all of his fellow soldiers. This final display of erudition by Benitez—when considered in conjunction with the other accomplishments of the students—symbolizes the recruits' mastery of the physical and intellectual demands of basic training.

If the army brass hats are astonished by the accomplishment of Rago's students, they are hardly disappointed with their results. When he initially conveys the colonel's ideas about the remediation program to Rago, Capt. Murdoch is unsure of the wisdom behind such an idea. However, he exhibits an unconditional confidence in Rago's ability, and he avoids monitoring Rago's daily lessons. Furthermore, Murdoch displays a sympathetic concern for the lack of academic preparedness of the Double Ds that typifies the film's intermingling of liberal and conservative positions. He expresses his lament to Rago thusly:

> I think the whole educational system is in deep trouble. You see, after their families let them down, they go to what's left of their schools. The teachers are underpaid, they're scared, so they stop teaching. But that doesn't stop them from handing out high school diplomas like they were toilet paper. Guess that means we gotta take them. The country better shape up soon, Bill . . . [Cause] we're going to hell in a handcart.

Rago has less interaction with Col. James, but he learns through Murdoch that the colonel is as sincere about helping the Double Ds "do pushups" as he is about giving them "a hand in the brain department." Importantly, both James and Murdoch are white.

On the other hand, Rago's encounters with Sgt. Cass, who is black, take on an extremely different tenor. Their meetings are quite contentious (but comically so) and become progressively more heated until Cass almost attacks Rago over the tardiness of two of Rago's students. One point of contention between the two men is that Cass does not share the same enthusiasm toward the usefulness of "book knowledge" as his superior officers. He, like Murdoch, believes that his trainees have poor comprehension skills, but Cass has no faith in the odd notion that a six-week English class will help. Once he discovers that Rago will be teaching Shakespeare to his recruits, Cass is skeptical about the class's benefit to the future soldiers. He is mildly humored by Rago's curriculum, but indicates that he expects the class not to interfere with his own training time.

What does it mean that *Renaissance Man*'s most prominent black authority figure has trouble appreciating the role that intellectual stimulation could play in the development of recruits into soldiers? Cass's attitude toward the remedial class appears even more odd when contrasted with the sincerity expressed about the educational welfare of the Double Ds by the white officers. As far as school films are concerned, the precedence discipline takes over classroom learning in Cass's mind recalls the model set forth by Joe Clark in *Lean on Me*. Whereas the other figures of authority are concerned with the Double D's lack of academic preparedness—all of whom are white—Cass has deemed their intellectual enrichment secondary to their development as soldiers. Interestingly, two film critics suggested in their respective reviews of the movie that the two actors who played Cass (Gregory Hines) and Murdoch (James Remar) were miscast and that both could have been more effective if their roles had been exchanged (Lowry 1994; Maslin 1994). The reasons behind the casting choices are unknown, but one thing is certain: had the roles been different, the racial politics would have been different as well.

Whitewashing the Military Image

It seems that *Renaissance Man* also makes a significant (unconscious though it may be) racial statement in its portrayal of the army as an institution, an institution that is one of the last preserves of white, male leadership. Throughout the movie, Sgt. Cass exemplifies the stern behavior stereotypically associated with military drill sergeants and with the military in general. He exhibits no tolerance for excuses, and expects superior performance from the recruits under his tutelage. In one scene, the sergeant reprimands Davis and Benitez for their tardiness. Rago attempts to cover for his students by taking the blame for their violation; his attempt merely incites the wrath of Cass, who comments on Rago's "smug, college-boy attitude" and accuses him of impugning his authority in the presence of his soldiers. Despite his stern behavior, it would be inaccurate to describe Cass as a mere ogre, for there are moments in the film when he is sensitive and encouraging to his trainees. Generally, though, he plays the role of a taskmaster. Hence, in his preoccupation with discipline, order, obedience, and punctuality, Cass embodies the traits commonly associated with military officers.

In contrast to Cass, Capt. Murdoch and Col. James are calm, understanding, and overly concerned with the well being of the recruits. Murdoch and James are neither the moronic nor the incompetent brass hats often featured in military comedies; far from it, they are capable, fully informed officers. *Renaissance Man*, then, takes the most unsavory characteristics of military personnel—inflexibility, a disregard for independent thinking—and requires its black authority figure to bear the brunt of these weighty stereotypes. Alternately, the film assigns the more pleasant traits to the white officers and cleanses them of the unpleasant details of the army, probably rendering the white men more likable, at least in the eyes of civilians. Thus, in an institution in which white men are overrepresented among the officer ranks relative to their numbers among all enlisted personnel (U.S. Bureau of the Census 1995), *Renaissance Man* rescues their image by making the lone African American carry the responsibility for its most unpalatable images.

Renaissance Man performs public relations work for the military—in the service of recuperating whiteness—in other clever ways.[13] As a citizen with no military experience, the film unsurprisingly sets Rago up to challenge the philosophies of the army from the moment he first arrives on the training camp. His discomfort with life on the base is depicted in several ways. For example, during Rago's only meeting with Col. James, Rago indirectly criticizes the military by recounting his experience marching in Washington, DC from Arlington Cemetery to the White House in order to protest American involvement in the Vietnam War. James counters by asking Rago if he remembers the name of the soldier he was honoring in the march. Rago does not remember, giving James (and by extension the military) the upper hand. Rago's antimilitary sentiment is therefore neutralized by James's response.

On a few occasions Rago registers his complaints with the military lifestyle with Sgt. Cass; in this sense, he serves as a foil to Cass. In most of the confrontations between the two men, Cass leaves the scene frustrated but certainly not the clear "victor" in the verbal battle as James is in the Rago–James discussion of Vietnam. However, in one particular encounter concerning punctuality, a prior subject of disagreement between the sergeant and the teacher, Cass *appears* to get the better of Rago. The teacher shows up late for class one day, and once the Double Ds find out that he was late because he was on a job interview, the insulted and upset students collectively storm out of the class. As Rago tries to apologize to his students, they ignore him, and he follows them (with great difficulty) along the "Victory Tower," their obstacle course. He ends up at the top with only one way down—to rappel a high wall—with Cass watching from below. In order to regain the respect of his students, Rago makes his embarrassing but successful descent and Cass walks away without a word. This scene accomplishes many goals for the film: Rago overcomes the obstacle to trust and communication between him and his students posed by the tunnel-visioned Cass; moreover, Rago, the white male intellectual, enters Cass's domain of black male physical prowess and conquers it, however clumsily; last, the movie supports the army's emphasis on personal responsibility by making Rago "learn a lesson" for being late, implicitly supporting the army's rules while silencing the person (i.e., Cass) who most symbolizes these rules.

As the film unfolds, it gradually offers more support for the army's philosophies while once again granting Rago a symbolic victory over Cass. Rago expends a significant amount of effort in the film locating the military records of Brian Davis's father. He finally recovers them and goes to deliver them to Davis by way of Cass one stormy evening when the soldiers are on bivouac. Rago tries to get Cass's attention, but he ignores him. Instead, Cass attempts to embarrass the Double Ds in front of their beloved teacher by ordering Melvin to recite something from Shakespeare. When Melvin is unsuccessful, Cass calls Benitez to do the job. Benitez surprises Cass (and perhaps everyone else) with a recitation of the King's "St. Crispin's Day" speech from *Henry V*. By having Benitez recite this speech, which includes the words, "We few, we happy few, we band of brothers," the filmmakers achieve two purposes. For one thing, it is yet another of the film's endorsements of the army—this time underscoring the importance of unit cohesion. For another, the scene counts as one more humiliation of Cass on his own turf at the hands of white men (Benitez and Rago). Cass has the unfortunate task of enduring this humiliating moment as he quietly commends Benitez and takes the records from Rago.

In short, the makers of *Renaissance Man* employ an interesting strategy in order to maintain the dignity of both whites and the military: they depict the white, male characters as sensitive, concerned, enlightened individuals, in definite opposition to military stereotypes; instead, they have the character of the black, male authority figure tote the traditional baggage of the military; also, they celebrate the traditional values of the military while stripping these values away from the black male, humiliating him, and demonstrating his unenlightened state; and they offer the military's values as solutions to the social ills that afflict its pariah-recruits. Through this clever strategy, the movie rehabilitates the much-maligned image of an institution dominated by white men without anyone getting hurt—that is, except for Sgt. Cass.

Renaissance Man as a Clinton-Era Text

A consideration of the profiles of the men and one woman who make up the Double Ds reveals one instance of the liberal/conservative duality that characterizes the film and exemplifies the lingering effects of the Reaganism in Clinton-era school films. Their individual profiles are based on familiar "types": the clueless Southern hillbilly; the semiliterate, former athlete; the class clown; the child of an irresponsible single parent, and so on. Together, the Double Ds are a group of potential soldiers who, in the words of Murdoch, need "a hand in the brain department." Their problem, however, as the audience (along with Rago) eventually learns is not that they are "intellectually challenged," but that they have not been challenged *enough* intellectually. The film spends a significant amount of time—more than most school films—delving into the backgrounds of *each* of the eight recruits. Its filmmakers seem to endorse the position that the students' academic underachievement is a result of their troubled lives, as opposed to the view that their lives are troubled because they are lacking in intellectual ability.

Yet the film's typical "happy ending" is not without misfortune. One of the Double Ds, Roosevelt Hobbs, demonstrates such an impressive comprehension of Shakespeare that Rago recommends that Murdoch investigate his background and consider reassigning him to an officer training program. It turns out that Hobbs is a crack dealer who is hiding out from the authorities under an alias; he is summarily arrested, of course, and carted off to prison. When Rago explains what happened to Hobbs's saddened classmates, they are inclined to blame Rago. But he refuses to accept the blame for Hobbs's imprisonment, and makes a moral pronouncement about the unfortunate incident with the help of Shakespeare: "The choices we make dictate the life we lead. 'To thine own self be true.' "

Why did the filmmakers choose to sacrifice one of the Double Ds? Perhaps to make the film more believable, since not everyone will be successful. But if verisimilitude were really the goal of the filmmakers, then why not have one of the white students be arrested for selling crack or some other drug, since the majority of drug users *and* sellers are white (Lusane and Desmond 1991; Moore 1995). Instead, the filmmakers rely on the easy stereotypical image of the black "street pharmacist." Further, the Hobbs incident is infused with the language of choice and personal responsibility so loved by Reagan, Bush, and their constituents.

One might accuse this critique of the Hobbs situation as unnecessarily harsh. After all, the filmmakers did portray Hobbs as the smartest student in Rago's class. Unfortunately, he was a smart student who made some bad choices for which he had to be punished. But all is not lost, because the film suggests that Hobbs may receive an earlier release from prison thanks to the benevolence of Rago and Murdoch. Even better is the fact that Rago's teaching left a meaningful impact on Hobbs's life: the audience learns through a letter to Rago that Hobbs borrows *Othello* from the prison library. Rago, then, is one of the endangered species of "old teachers who loved Shakespeare or Austen or Donne" (Bloom 1987, 65) whose breeding of a new flock of Shakespeare enthusiasts will ensure the perpetuation of the species.

Although the filmmakers present a best-case scenario of Hobbs's situation, one wonders why the situation had to appear in the film at all. If making Hobbs a smart but criminal character was an attempt to convince viewers that people who make mistakes should be given second chances, then that convincing could have been accomplished without resorting to racial stereotyping. The more salient lesson this incident seems to teach is actually related to Rago's tardiness episode earlier in the film: that people must suffer the consequences of their actions. In yet another instance, the underlying message of *Renaissance Man* shows itself to be merely a wolf of traditionalism dressed in sheep's clothing. The episode involving Hobbs is the epitome of what is at work throughout the film, and that is the infiltration of an ostensibly liberal racial agenda by the legacy of the Reagan-era politics. Let us now consider how a similar agenda operates in *Dangerous Minds*, a more recent nineties school film.

The Minds behind a Dangerous Film

Dangerous Minds got off to a wonderful start right after its release. It was a two-time winner in the box-office race and the song "Gangsta's Paradise" from the film's soundtrack was selling well. Undoubtedly, the appearance of Michelle Pfeiffer in the film's central role and the chart-topping popularity of the first single from its soundtrack must be factored into the audience appeal equation. But I realized after viewing the film that its allure lay in the literal and figurative scripts from which it operates. Although the plot of *Dangerous Minds* is merely an updated version of a tale that has been told on film many times, it is nonetheless an interesting story, particularly since it was inspired by real-life experiences. Behind the script's "success against all odds" story line and façade of multicultural inclusivity, however, lies a reactionary and, indeed, "dangerous" narrative. This narrative attempts to paint a simplistic picture of society where the incorrigibility of black and Latino youths is the problem to be solved, black adults are the main obstacle, and whites are the solution. Not a surprising portrait of America, given the climate of white backlash and white panic that existed at the time.

The narrative of *Dangerous Minds* is intimately connected to this climate of white backlash. Take, for example, the change of the book's title from *My Posse Don't Do Homework* to its filmic title *Dangerous Minds*, which is in effect a change from a real quote from the author's former student to a label that reinforces the myth that these students pose a threat to society. In fact, I see the narrative as another expression of the desire to regain a romanticized past of relatively uncontested white dominance. And the film *Dangerous Minds* fits in perfectly with the pattern of practices intended to secure this dominance. I examine the production of *Dangerous Minds* in order to demonstrate why it should be seen as an apparatus for the recovery of former racial positionalities in America and how it attempts to accomplish its underlying goals.

Dangerous Minds is based on the real-life story of LouAnne Johnson, a white ex-marine who was assigned on her first teaching assignment to a group of capable but troubled students. In the northern California public school where she began her teaching career, Johnson encountered these students in the "Academy," a school-within-a-school for mostly bussed white, black, and Latino youths with academic difficulties. The film essentially portrays the trials and tribulations faced by Johnson and her students as recounted in her book, *My Posse Don't Do Homework*.

For the most part, the film's plot revolves around "slices of life" from Johnson's classroom. For example, the film follows Callie, a black, female student who is depicted as perceptive, mature, and dedicated. Her progress goes relatively undisturbed until the school's administrators become aware that she is pregnant and inform her that she should transfer to another school. Raul, another of Johnson's hardest-working students, finds himself and a male friend in a brawl with a third male student (all three are presumably Chicano) and ends up suspended from school. Later in the film, Raul wins a class contest and earns a free dinner with Johnson at an upscale restaurant. In order to obtain a nice jacket to wear to dinner, it is implied that he obtains the money for the jacket by doing something illegal. Johnson responds by loaning him the money to pay for it and making a deal with him to repay her by the time he graduates. The film also focuses on Emilio, the third student from the fight, who is initially portrayed as the invincible class "leader" but is finally forced to confront his fears of reading and of his involvement with an ex-convict.

Despite these seemingly student-centered anecdotes, *Dangerous Minds* is proudly concerned with tracing Johnson's experiences as a teacher at Carlmont High (or in Disney's version, Parkmont High). The film's central focus, though, is on Johnson's ability to connect with her students and help them to learn and to graduate, thanks to her use of tactics derived from boot camp, her penchant for giving unorthodox rewards, her willingness to get to know her students and their families, and her capacity for love and compassion. The film therefore deals with Johnson's contributions to each of her students' lives and her opposition to the school's rigid and oppressive administrative policies. Thus, *Dangerous Minds*, in its depiction of these elements of her experiences, is a success story.

In contrast to the film's message of uplift, however, is the fact that its racial representations are regressive and demeaning and the fact that its story line is disturbingly detached from the larger sociocultural context. In her insightfully critical review of the film, Amy Taubin (1995) writes that "*Dangerous Minds* is, in all ways, so removed from reality, that it's not worth getting upset about" (52). Her comment is almost convincing. But once the film is contextualized within the agenda of the film's producers, within Disney's pattern of pseudo-innocent films, within the nexus of dominant Hollywood ideologies, and, finally, within the 1990s realization of the cinema of recuperation, the film's true colors are revealed and it becomes extremely upsetting.

Disney's Dangerous Minds: *A Hollywood Success Story*

Much has been made of the unexpected financial success of *Dangerous Minds* since its release in August of 1995.

The film opened at the top of the domestic box office making $14.9 million in its first weekend, remained the number one film for two weeks, and earned over $84 million by the end of 1995.[14] Its attractiveness to viewers has been attributed in large part to Michelle Pfeiffer, who has the starring role, as well as to its hip-hop soundtrack, which has sold over 3 million copies.[15] After the film's first weekend of release, for example, producer Jerry Bruckheimer was quoted as saying, "A star and music are driving the picture" (Green 1995). It seems, though, that the film's appeal is explainable in another way: it fits neatly and securely onto Disney's shelf of "great white hope" films, it continues Hollywood's filmic typology, and it provides an accessible narrative for the recovery of white racial dominance.

Given the involvement of the powerhouse production team of Jerry Bruckheimer and Don Simpson—responsible in the past for highly popular films such as *Top Gun* (1986) and *Beverly Hills Cop II* (1987)—*Dangerous Minds* was undoubtedly destined, or at least intended, to become a box-office success. Coming off the heels of recent successes like *Bad Boys* (1995) and *Crimson Tide* (1995), films that caught the attention of black viewers, Simpson and Bruckheimer surely wanted another popular production, one that would still appeal to African Americans while attracting a larger white audience as well as other nonwhites. Perhaps they also wanted to inject an oft-told story with new vigor. In short, they needed a gimmick, or several of them; hence, the choice of Hollywood superstar Pfeiffer for the leading role, the recruitment of well-liked hip hop and R&B artists for the film's soundtrack album, and the hiring of prolific screenwriter Ronald Bass. These elements, however, were evidently not sufficient for the filmmakers. Not content to portray Johnson's story on film as it was recounted in her book and interviews, Bruckheimer, Simpson, and Bass intentionally fictionalized her experiences. These three agents flipped the script, so to speak, apparently not only to guarantee the film's financial success, but to further their racial politics.

The creation of *Dangerous Minds* was also informed by the politics of Disney. As in *Dead Poets Society*, Disney's influence in *Dangerous Minds* hides behind the film's family-oriented, feel-good story line. The film is a fitting addition to Disney's collection of filmic texts because it carries on the company's venerable tradition of depicting its heroes as white, middle-class, and, often times, male. There are several films during that period that exemplify Disney's version of the world: *Good Morning, Vietnam*, *Pretty Woman* (1990), *Beauty and the Beast* (1991), and *The Scarlet Letter* (1995). While there are admittedly qualitative differences among the above films, what is common among them is the promotion of cultural traits, perspectives, and values associated with the professional and upper classes, with white Europeans and Americans, and with the demonization of the cultural expressions of marginalized groups.

If *Dangerous Minds* does not paradigmatically disrupt the flow of texts from Disney's production houses or from the imaginations of Simpson and Bruckheimer, then neither does it have difficulty finding its way into the ideological mainstream of Hollywood's corpus of movies. Even a cursory review of a list of mainstream Hollywood films can lead to the categorization of these films under the rubric white (male) savior films. Although this category obviously does not account for every Hollywood release, it does encompass the majority of its mainstream, in which many of Disney's films can easily be included. A few examples of feature films from the recent past should help to underscore the argument. Among the examples are several critical and financial successes, such as *Glory* (1989), *Terminator 2: Judgment Day* (1991), *Rising Sun* (1993), and *Forrest Gump* (1994), as well as lesser-seen films such as *The Journey of August King* (1995).[16]

Although one may argue that my typology collapses the complexity of the films I attempt to categorize, I believe that *Dangerous Minds* is incontestably a white savior film. Its tale of a white teacher who tames and rescues a group of bright but wayward youth, wrapped in an attention-grabbing bow and sealed with a happy ending, immediately qualifies it for membership in this filmic category. In this way, it also maintains the legacy of 1980s school films while sporting a

Clinton-era facelift. *Dangerous Minds* is also in keeping with a series of school films that tackle the same subject matter, such as *Blackboard Jungle*, *To Sir, with Love*, and *Conrack*. Thus, in its elevation of whiteness and its use of a tried-and-true story line, the film, like *Dead Poets Society*, is definitely mainstream and typically Disney. This assertion is perhaps so obvious that it seems unnecessary to state. The point, then, of naming *Dangerous Minds* as traditional is to emphasize that it does not do much, technically speaking, that is new. Yet it includes enough new elements to reinvigorate its story, make it relevant to today's context, and buttress its racial agenda. Disney's release of this film acts as an attempt to reclaim the privileged position of whiteness, a position that is being increasingly threatened at every turn. Put another way, the film's traditionalism can be seen as a nostalgic yearning for a romanticized past of white dominance.

Recovering White Racial Privilege through the Film

Dangerous Minds attempts to reestablish white supremacy by presenting its representations as "the way things are." In an effort to make these representations seem natural, the film tries to disguise the ideologies that inform it and consequently invites its audience to accept the narrative it offers as real. The film is, after all, based on a true account, so the filmmakers can expect many viewers to believe in the narrative. But *Dangerous Minds* has been deliberately severed from the real experiences that gave birth to it. Given this severing, one might wonder how Johnson's meaningful story has been transformed into an apparatus for reasserting the supremacy of whiteness. In order to understand this transformation, it will be necessary to scrutinize the film's characters, its decontextualized narrative, and the plot changes brought about by the filmmakers.

Aside from the teacher and three students mentioned above, there are three other central characters in *Dangerous Minds*: the school's principal, the assistant principal, and LouAnne Johnson's friend and fellow teacher. Johnson's character, as I have already stated, is the focal point of the film. Her character is portrayed as patient, sensitive to the needs and feelings of her students, silent yet strong, and flexible. Another trait is her benevolence, as demonstrated by her donations of money, time, energy, and compassion. Other salient characteristics are her divorced status and her background as a marine. The principal, as a black man, is shown as the antithesis of Johnson's character. He is ineffective in dealing with students in the Academy, largely due to his cold and insensitive attitude toward teachers and students alike. He is overly concerned with rules and procedures in an environment that begs for constant innovation and exception-making. The principal also acts as a barrier to Johnson's successful pedagogical practices, and his rigidity is responsible for the death of Johnson's student.

As we might expect, the assistant principal—who, like Johnson, is white and female—has the task of doing the principal's dirty work; it is she who informs Johnson of the unstated school rules. Unlike the other administrator and Johnson she appears to be lacking in conviction and authority; her character comes across as sheepish and helpless. She seems torn between being slavishly attached to the rules and actually being guided by compassion. The last of the adult bunch is Johnson's friend and colleague, who has been teaching at the high school for a while prior to Johnson's arrival. We learn that he was a friend of hers and of her ex-husband and was the one who helped her to land the teaching position. He does not appear very often in the film, but when he does, he is seen as a confidant to and source of motivation to Johnson. She relies to some degree on him for personal advice and pedagogical guidance. And he is white.

I contend that the racial (and gender) identities of these characters, along with their personalities and backgrounds and their roles in the story, speak volumes about the agenda of the film.[17] To begin with, Johnson is depicted as the teacher-savior who helps to turn around the lives of a group of academically unsuccessful, mostly nonwhite youths. These youths consist of the stereotypical pregnant black female student (Callie), and the 1990s versions of the bandido/buffoon Raul and the bandido/Latin lover Emilio (Pettit 1980; Berg 1990).

Her benevolence comes across through her acts of loaning money, providing refuge, and donating dinners to her students. These acts also highlight their dependence on her and can be read as a subtle suggestion that nonwhites "owe" whites for their advancement. Finally, this salvific image is reiterated at the end of the film when the students tell Johnson that she is their "light."

In sharp contrast, the principal appears as the enemy of the educational process. His punctiliousness stands in the way of Johnson's more sensitive and adaptational approach, and in the end her manner proves to be more effective. The black principal, unable to tame his own kind, stands in contradistinction to the white teacher-savior, who becomes the heroine of the white privilege rescue attempt. The principal may also be compared to Sgt. Cass of *Renaissance Man*, a man similarly obsessed with rules, and contrasted with the black principal in *Lean on Me*, who *is* able to control his students, but only with a tough-love approach that involves carrying around a bat through the school. Compare these three black educators—one, an ineffectual bureaucrat, one, an effective but ultimately impugned taskmaster, and the third a boisterous, bat-toting "brute"—to Johnson, who of course uses a more "civilized" approach to help her students out of their plight. Moreover, the emphasis on Johnson's background in the Marine Corps is another piece of the recuperation puzzle. Her creative use of military-based disciplinary tactics in her classroom legitimates, like *Renaissance Man*, one of the last bastions of white male rule (Colin Powell's former head position notwithstanding).

Johnson's teacher friend is relatively absent on screen throughout the film, yet he is in some ways omnipresent. As her closest colleague, he is familiar with Johnson's personal life and work life. He contributes to her decision to remain at the school, and his occasional advice helps her to cope with her daily frustrations and challenges. It is also important to note that his class of students in the film is more quiet, more cooperative, and comprised of more whites than Johnson's class, implying that their whiteness (and perhaps his maleness) is the reason for their "better" behavior. There appears to be, then, an erasure of his physical presence, but a ubiquitousness of his influence. As for the assistant principal, the other administrator in the film, her sheepishness and lack of contact with and control over the students in the Academy underscores Johnson's role as their heroine.

One of Johnson's first statements to her students after entering her new school is, "There are no victims here." In numerous other scenes, she emphasizes their capacity to make choices. While she likely means to convey to them her intention to treat them as equals in her classroom, these statements carry other potential meanings for the audience; that is, they may invoke the conservative discourses of individual choice and responsibility, discourses that even Clinton used often to his political advantage (Williams 1998). These discourses are related to what Gresson (1995) calls "rhetorical reversal." Reversal, he writes, "pertains to the power to name, define, and negotiate reality" (145). This happens when people who issue accusations of racism are considered by others (usually those on the Right) to be part of the "cult of victimization." And the ones with the power to define reality usually occupy a privileged position in relation to those whose situation they are defining. *Dangerous Minds* might have followed the lead of *Boyz N the Hood* (1991), a film that offered a more responsible and contextualized understanding of individual choice. Dyson (1992) notes that the director of *Boyz N the Hood*, John Singleton, obviously understands that "character is not only structured by the choices we make, but by the range of choices we have to choose from—choices for which individuals alone are not responsible" (135).

Based on Johnson's choice of words, one would guess that her "victim" statement was delivered to her students in anticipation of them offering a defense of their "at-risk" conditions. Therefore, to counteract any possible pleas for understanding like "I'm in the Academy because Mexican Americans are discriminated against" or "I can't concentrate because I didn't have money for breakfast," she "reversed" the situation by claiming that "there are no victims" in her classroom. If she were able to isolate her classroom from structural forms of discrimination and asymmetries of power, her ideal environment might be attainable. We know, however, that this is not possible, and it is the decision to include this statement in the film and the film's decontextualized nature that are troubling.

A few writers have commented on the film's uprooting of Johnson's story from its societal context. Todd McCarthy (1995), for instance, declares the following:

> There's no portrait of the kids as part of a larger culture, no sense of other forces—inspirational or oppressive—working on their lives, no discussion of the currents in society that inform the attitudes of the students, no presentation of their personal lives, emotions or interests. (55)

I think this is more of a shortcoming of the film than of Johnson's story, because the real teacher seems aware of the importance of seeing students' difficulties within their proper context. For example, during a television talk show she asks, "How can you teach people if you can't understand them?" Later in the show she makes another pertinent comment: "If we want [kids] to be responsible, people with discipline, [adults] have to have our own discipline, and we have to teach them."[18]

However, the filmmakers fail to depict the crucial connection between the students' classroom behavior and the challenges they may face, such as poverty, racism, and learning disabilities. From this severely decontextualized perspective in which power is erased, it is a short step to the discourse of "individual responsibility." For to ignore the workings of power and to feign ignorance of privilege is to narratively reverse the existing situation and hold individuals responsible for their positions in society regardless of the oppression they face. In sum, the film's racial politics are made manifest through its avoidance of cultural context.

There are several other tactics that the film's producers employ that significantly alter and undermine the original story. As stated above, the large majority of Johnson's class in the film consisted of Latino and black students. But in the actual Academy at Carlmont, one-third of the students were white. Johnson speculated that the filmmakers changed the racial composition of the class because they "were in love with the idea of Michelle [Pfeiffer] being White and the kids not . . . They have this notion that school is only a problem for poor minority kids" (Britt 1995). Another example occurs in the scene involving a black grandmother who spurns Johnson's teacherly concern for the woman's grandsons by saying, "Find yourself some other poor boys to save." In actuality, the boys' grandmother had her sons leave school to help run the family business. Johnson later received a rose from one of the boys and a card that read, "I am still reading poetry, and I will always think of you as my teacher." Apparently the real Johnson found this scene "insulting," and said that "it taps into a misconception . . . that black parents are going to be antagonistic toward white people, which wasn't my experience" (Britt 1995). Additionally, Johnson's character in *Dangerous Minds* uses Dylan Thomas and Bob Dylan to teach her students poetry. In the book, however, she uses rap music to get more difficult students involved. This substitution served to change her pedagogy from something that worked to something the filmmakers obviously wanted to see take place.

To add injury to insult, the filmmakers tried to add elements to the teacher's past that hadn't existed in her real life, ostensibly to render the film more appealing to viewers. As journalist John Scalzi (1995) reports, Johnson had the following to say about the filming experience: "I think the final version of the film was much closer to the truth than it was originally. For example, in the first script they have me as a single mother with a seven-year-old daughter." Johnson successfully protested this modification, which was possibly an attempt to portray her as a superteacher *and* a mother, thus reinforcing her heroism. What she was unable to prevent, at least initially, was the addition of a male love interest for her character. According to her, the producers believed that the "glamorous" Pfeiffer needed a boyfriend. When she disagreed, they responded by saying, "No, you don't understand, this is Hollywood, and [audiences] need a love interest," and proceeded to make the addition (Scalzi 1995). Luckily, preview audiences reacted negatively to the added boyfriend, and the filmmakers removed the implicated scenes, thus demonstrating the disparity between what the producers wanted to show and what viewers wanted to see.[19]

But why did Disney's filmmakers insist on tampering with the story of *My Posse Don't Do Homework* so much? One could argue that the goal of their narrative sabotage was to make the film more palatable to audiences and ultimately to sell more tickets. This is, of course, a common practice in translating books to the silver screen. It seems, though, that the actual story of Johnson's triumphs and failures at Carlmont High School was dynamic enough without the producers' editorial changes. If they had indeed meant to fictionalize Johnson's story as is implied by the change of the school's name and the addition of a romantic subplot, they could have altered several other elements without diminishing the story's power—like substituting a black or Latino teacher for Johnson or making all the students white. This, however, would have impeded their apparent racial agenda.

The aforementioned examples make a powerful statement about the filmmakers' attempt to construct a story that was not corroborated by the reality portrayed in Johnson's accounts. The fact that the filmmakers "colored" Johnson's classroom, sensationalized a few episodes, sanitized her curriculum, and essentially lied in a "true" story has to be seen as not just creative filmmaking, but as an intentional act of racial and cultural polarization under the guise of multicultural appeal. Contrary to the message the film tries to convey, it is not the "minds" of Johnson's students that are dangerous, but the agenda of the filmmakers and the message of the film itself. In sum, the demonization of nonwhites, separation from the social/political/cultural context, and racial and class polarization present in *Dangerous Minds* create the impression that the "White man's burden" of having to civilize the Other will never end.

Civilization through Music

The most recent of the Clinton-era school films and the last one to be analyzed here is *Music of the Heart*. The film, an

Oscar-nominated Miramax production, was released in October of 1999—well after the summer blockbuster season but early enough in the year to qualify for Academy Award consideration, and with enough stars like Meryl Streep, Angela Bassett, and musician Gloria Estefan to attract the attention of both movie reviewers and general audiences. An additional appealing fact was that the film was directed by Wes Craven, who had become (in)famous for an almost 30-year-long career of making horror films such as *The Last House on the Left* (1972), *A Nightmare on Elm Street* (1984), and *Scream* (1996) and *Scream 2* (1997). In fact, *Music of the Heart* was the first "non-genre" film he had ever had the opportunity to direct (Malanowski 1999).

The plot of the movie is by now a familiar one to viewers who have seen many of the school films that preceded it. It is a dramatized account, inspired by an earlier documentary, of a real-life educator's experiences teaching music in a New York elementary school. In 1980, after finding herself suddenly alone with two boys to raise, the educator, Roberta Guaspari, talked her way into a job as a violin instructor at an East Harlem elementary school.[20] The principal was initially reluctant to take her on, but an impromptu demonstration of Guaspari's success with teaching her own sons helped to convince the principal to hire her on a conditional basis. Guaspari brought her talent, her energy, and her 50 violins that she had bought during her marriage to a classroom of students who were wholly unaccustomed to the instrument.

Despite facing the initial chaos of her classroom, the lack of interest and commitment from some of her students, and the unsupportive attitudes of a few of her fellow teachers—especially the particularly jaded music teacher—Guaspari successfully reaches and teaches her young charges. The movie paints a generally positive picture of her experiences, missteps and all, depicted most clearly when it shows Guaspari ten years after her debut running violin programs at three different schools and being interviewed by a photojournalist for an article the journalist is writing. All of a sudden, the board of education cuts the funding for arts and music instruction and Guaspari finds herself with a jobless future. In response, the resourceful educator organizes a charity concert to raise money to support her programs and herself. By drawing on the assistance of the photojournalist and of her current romantic interest, she ultimately lands Carnegie Hall as a venue and, as the icing on the cake, gets some famous violinists like Itzhak Perlman, Issac Stern, and Arnold Steinhardt to perform with her and her students. Guaspari's efforts are an amazing success, thereby saving her violin programs from elimination.

Different City, Similar Narrative

The preceding synopsis certainly resembles the plots of several earlier school films—perhaps most closely that of *Dangerous Minds*. Unsurprisingly, many popular reviewers compared (either favorably or unfavorably) *Music of the Heart* with a few of the earlier films, *Dangerous Minds* being one of the most frequently cited ones.[21] This is understandable, because there are in fact numerous apparent similarities between the two. For one thing, both films are based on real people and their real experiences, and are inspired (at least in part) by true accounts (one an autobiography, one a documentary) of those experiences. In addition, the central educator figure in each is a woman—a divorced single woman at that, which is atypical among the school films from the 1980s and 1990s (and among the entire subgenre). Both women are inexperienced teachers when the films begin, and both are white teachers in multiracial public schools and, specifically, in predominantly nonwhite classrooms. Moreover, the schools are located in urban settings that, in the films, are marked by violence. During their experiences, each of them bumps up against other uncaring educators, faces obstructionist or unconcerned parents, and is threatened by the vagaries of school systems that are portrayed as unnecessarily rigid, insensitive, or wrong-headed. Finally, despite their occasional mistakes, both of them eventually succeed through a mixture of tough love, creativity, and tireless dedication.

In short, *Music of the Heart* is very much in keeping with the patterns established by the school films discussed previously. The film follows the path already created by the subgenre of movies from the 1980s and 1990s. It features an individual educator-hero who succeeds primarily by his or her own devices; a bureaucratic school system that seems to abide or reward failure (through tenure for instance) and to suppress real concern, devotion, and talent while forgetting what meaningful education is really about; an emphasis on the dangers of the inner city; and a triumphant narrative. It also treads in the particular footsteps of *Dangerous Minds*, with its (neo)liberal racial and spatial politics embedded within its humanist, multicultural narrative and its alteration of certain real-life events to tell a different story. There are in fact enough consistencies between *Music of the Heart* and the other school films examined in this chapter that it can also be understood as a Clinton-era text. The following sections attempt to flesh out just how this is so.

The Genesis of Music of the Heart

The process by which the film came to be made, and the people who were a part of it, make for an interesting tale and are helpful in understanding the movie's intricacies. The director, Wes Craven, never intended to focus his career on horror films. But his history and success in the genre worked against him when it came to producers considering him for other kinds of projects. Ultimately, however, it was the success of *Scream*—a $103 million domestic box-office hit that creatively toyed with many of the conventions of the horror genre—that helped him to escape from the box in which he found himself. During a celebratory dinner, the founders and chiefs of Miramax, Bob and Harvey Weinstein, made him an irresistible offer: a three-picture deal, one of which could be any non-horror film on the Miramax development schedule (Malanowski 1999).[22]

Craven selected *Fifty Violins*—the name would later be changed due to audience testing. The choice was serendipitous: Craven had earlier served on the Academy Awards prescreening committee for the documentary category, and one of the films he saw was *Small Wonders* (1995), the true story of Roberta Guaspari that inspired the making of *Music of the Heart*. Craven had also been a humanities instructor and was a fan of classical music, so he felt connected to the project for many reasons. Even though Craven's film was inspired by the documentary, it was not designed to be a direct adaptation of the documentary's content. In order to help create the dramatized version of Guaspari's experiences, the film's screenwriter spent a great deal of time with Guaspari (Malanowski 1999) and Guaspari also served as an on-site consultant to the film (Kwiatkowski 1999).

Despite the limited appeal that one might imagine a story about kids and violins would have and despite Craven's inexperience making films outside of the horror genre, *Music of the Heart* was not being treated as some art house experiment or mere gesture of indulgence toward a moneymaking director. The choices made by the filmmakers suggest that they had more in mind for the film than that. For instance, Miramax initially urged Craven to cast popular performer Madonna in the lead role. They worked together for a while, but the studio and the performer eventually parted ways because of "creative differences" (Stein 1999). After searching unsuccessfully for replacements such as Susan Sarandon and Julia Roberts, Craven was finally able to attract the highly regarded Streep on his second try, just two days before having to vacate the production's office space (Malanowski 1999). And casting Angela Bassett and Gloria Estefan only added to the film's star appeal. In addition, the documentary that had inspired the movie project had been nominated for an Oscar in 1996; this pedigree was undoubtedly helpful in promoting the film. Furthermore, although the Weinsteins had earned a reputation for taking risks on cutting-edge films and sponsoring breakout directors,[23] their studio had by 1999 entered what Peter Biskind calls its "third phase" during which it surpassed its competition and "enjoyed enormous critical goodwill and commercial success" (Biskind 2004, 376). Last, the title song on the soundtrack was written by the prolific and prosperous Diane Warren and performed by pop group *NSYNC and Estefan.

In short, by all appearances *Music of the Heart* was designed to fit well within Hollywood's mainstream, with no indication that the film intended to challenge the conventions or messages of the industry. Its origins, casting choices, and production history create an impression of a film created for maximum, not minimum, appeal. And based on a few key indicators, it appears that the filmmakers were successful. The film's take of more than $14 million at the domestic box office is an unremarkable figure. The movie was recognized, however, by a Golden Globe nomination for actress Streep, as well as Oscar nominations for Streep and for Diane Warren in the original song category. Warren's song was also nominated for a Grammy Award. Moreover, around the time of the movie's release the song reached the number two position on the Billboard singles chart and actually topped the singles sales chart. Given the film's status as a mainstream product, let us examine whether we can describe its dominant messages as also being mainstream.

Challenging the System and Departing from the Norm—Almost

In an earlier section I enumerated several of the ways in which *Music of the Heart* resembled *Dangerous Minds* and some of the other school films. Still, the narrative of the more recent film diverges in some interesting ways from its counterparts, and these divergences are worth exploring in order to assess the degree to which the film as a whole exhibits their familiar ideological patterns. *Music of the Heart* begins at the beginning of its central character's teaching career. She is not depicted, at least initially, as some superteacher transferred into a new environment or a maverick educator ready to shake an entire school out of its doldrums. Instead, Guaspari is a mother in need of a job, and she creates an opportunity for herself out of a mix of necessity, passion, and persistence. Her initial experiences with her new students show her to be uncertain and a bit lacking in her ability to manage the classroom environment. Soon enough, though, she gets her footing and begins to reach and teach her students.

This does not mean that Guaspari's professional life is without obstacles, however. Although her principal is generally supportive, Guaspari faces opposition in the form of a cranky, pessimistic, burned-out teacher named Dennis Raush; in fact, he is the only full-time music teacher at the school. Raush's jaded perspective concerning the students has apparently developed over his lengthy tenure at the school. While in the principal's office just after Guaspari has "auditioned" with the principal for her position, he condescends to tell her, "I think I know these students . . . Their attention span doesn't go past do, re, mi. You know, maybe on a good day I can get them to fa." In contrast to Raush, Guaspari displays an amazing dedication to her work throughout the film and a consistent belief in her students' capacity to learn. Indeed, the audacious act of thinking that urban students could learn to play an instrument as difficult as the violin is juxtaposed against Raush's extremely limited view of what those students could accomplish. Raush resurfaces several times during the movie, each time serving as the embodiment of uncaring, ineffective teaching—and each time, Guaspari's character challenges what he represents.

Furthermore, the film's characterization of Guaspari is not that of an angel. She is, for example, often commanding and unyielding. Several students find themselves on the receiving end of her blunt criticisms and matter-of-fact dismissals. Indeed, over time Guaspari becomes known for her tough love approach to teaching music. On occasion her approach crosses the line into insensitivity, like when she reprimands a student who is wearing leg braces and another whose grandmother died, both facts unbeknownst to her. Nevertheless, when her harshness is misplaced, it always turns into compassion. Moreover, the belief that she has the best interests of the students at heart is never questioned in the film's narrative. It is Guaspari's no-excuses approach, in fact, that helps her to get her students prepared for the charity concert that functions as a part of the film's denouement. All in all, despite her unpalatable qualities, Guaspari is offered as a refreshingly talented educator who through a mixture of hard work and uncompromising standards infuses a body of indifferent teachers and unchallenged students with a spirit of success.

Guaspari's experiences constitute an amazing story, and the impact and reach of her work should not be diminished. But a few contrasts between her real life and the way in which the film portrays it reveal some important details that beg for deeper exploration. *Music of the Heart*, like *Lean on Me* and *Dangerous Minds*, owes its existence to real-life events and actually advertises that it is based on a real-life story. And again, as with the earlier films, the fictionalized, big-screen version offers some interesting alterations of the true accounts (which allow it to send some different messages). One of those most significant alterations made by the film involves the changes that it made to the identity of the school where much of the action takes place. In fact, this modification of the school's character was the topic of discussion in several media stories. The school, named in the movie as Central Park East I Elementary (CPE), was the actual school where the real-life Guaspari worked. The actual school, however, was no mere educational wasteland or den of mediocrity. By the time Guaspari arrived to seek a job at CPE in 1980, the school had been open for years, founded in 1974 with Deborah Meier as its founding director, as part of a network of alternative public schools in East Harlem. The schools were notable for many reasons, one of which is that they were among the first institutions in the "small schools" movement (Meier 2000). The school was and is known (even at the time of Guaspari's time there) as a progressive institution dedicated to a model of school reform based on substantive staff involvement and meaningful community connections (Bensman 2000). In fact, Meier received a prestigious MacArthur Fellowship in 1987 because of her work as "teacher-director" of CPE. Interestingly, these details suggest the existence of a very different school than the one projected onto the screen.

This alternative depiction was the source of much irritation among some current and former educators at CPE. The director of the school at the time of the movie's release complained about the depiction on behalf of the teachers in her school.

Although she herself had not seen it, a wire story in the *Montreal Gazette* (1999) reported that others lamented the representation of the school's neighborhood as "very dangerous" and argued that an apology for the representation should be added to the film before the opening credits. In a similar way, another one of the school's educators contested part of the story told in *Music of the Heart*. After the film appeared in theaters, the regular music teacher of CPE, Barry Solowey, wrote of his rich experiences teaching recorder and being involved in the music program at the school as a way of demonstrating how these experiences were at odds with what viewers were exposed to in the film. He contends that the value of the music and arts programs was maligned in the film and expresses dissatisfaction about "the misrepresentation of what has been not only a wonderful music program . . . but an overall successful school that sees the arts as essential to every child's education" (Solowey 2000, 11).

But the CPE staff member whose public criticisms were most extensive and most prominent is Meier. In one such example, Meier wrote at length about the school's past, her experiences there, and the ways in which the film's portrayal diverged significantly from the real history of the school—and even from the account told in *Small Wonders*. She details the ways in which the film serves as accurate representation of Guaspari's tenure at CPE, and then goes on to critique how the film's "creative take on reality" results in a modified—and problematic—representation of Guaspari's story. For instance, as opposed to the film's presentation of a principal and violin teacher who collaborate in the midst of a body of indifferent or obstructionist teachers "to demonstrate what can be done for children" with caring, toughness, and high expectations, the real CPE was "a largely staff-run school" whose capacity to successfully educate so many students was based on "collective strength and sense of mutual responsibility to one another and the community" (Meier 2000, 64). Meier even praises Solowey for being an important piece of CPE's success through his 25-plus years of tireless, proficient work with all of the school's students, quite the opposite of the "mean and lazy" music teacher constructed in the movie.

Through these and several other fictionalizations, she argues that the film did damage to the fine, hard-earned reputation of the school. "In short," Meier writes, Guaspari "was one star among many," and not the sole bright light and savior of CPE.

This aspect of *Music of the Heart* begs a familiar question: why the modifications—or, better yet, why these modifications? It is quite understandable that the filmmakers would want to create as much dramatic tension as possible in order to make their film more appealing to audiences. And by setting up the central character as a "lone ranger in a hostile environment," in the words of one writer (Hartocollis 1999), the filmmakers presumably believed that their main character would seem that much more heroic. On an even grander scale, Guaspari's character may have even been envisioned by the makers of the film as a challenge to public schooling in general. Perhaps Guaspari was intended to be an example of what it takes to succeed against all odds, to overcome all that is wrong with public education. This, of course, is all speculative. However, what is clear is that, while the movie's narrative may have been designed to warm the heart and to champion talented educators, it fits securely within the pattern among school films of creating individual educators as heroes and overlooks the importance of having effective, supportive institutions that enable these educators to succeed.

How *Music of the Heart* handles race is another area in which the film's fictionalized retelling matters a great deal. One on level, its narrative seems rather progressive in the messages that it sends about race. First of all, almost all of the students in the school are racial minorities; the faces that we see in the hallways and in Guaspari's classroom are predominantly African American and Latino.[24] Yet unlike the other school films with multiracial student bodies, the students here are generally not portrayed as troubled, delinquent, violent, or incapable of learning. The film does depict them as unchallenged to some degree, but it roots this problem in the institution, not in the individuals themselves. As a more specific example, the character of the principal, who was fictionalized for the film, is an African American woman named Janet Williams. This is an interesting casting choice since Meier, the principal (or "teacher-director" as the position was defined then) of CPE while Guaspari was there, was a white woman. The reason for this characterization is unclear, but since Williams is portrayed as one the few educators who was open to Guaspari's work at the school, the characterization is what we might think of as a favorable one. The principal is painted as tough, for sure, but she is the one who agreed to take a chance on Guaspari by allowing her to teach at the school and remains supportive throughout the film. This portrait stands at odds with the black male principal in *Dangerous Minds*, suggesting that a different racial agenda may be at work in *Music of the Heart*. Hence, the film casts Williams in a positive light.

It does the same thing with another prominent nonwhite character, Isabel Vasquez, one of Guaspari's fellow teachers. Vasquez first befriends Guaspari in the school's courtyard. Guaspari has just sat down to have lunch when she sees two teachers passing by. She cordially offers them a seat, but they rebuff her by refusing to acknowledge her presence. One of the women (both African American) identifies Guaspari to the other as the new substitute teacher; the second woman's response makes it clear that they had discussed Guaspari before, and not too favorably. Vasquez walks up and facetiously says, "What's the matter, you got cooties or somethin'?" When Guaspari wonders aloud why the teachers seem to hate her, Vasquez explains their response as less about dislike than about reflexive indifference toward teachers of "special programs" like the violin classes. Vasquez does say, however, that one of the impolite African American women is also a "bitch." She admits to having "ulterior motives" behind being friendly—wanting her daughter in Guaspari's class the next year—but the movie suggests that her supportive attitude transcends any selfish desires she might have. It bears this out throughout the entire story as Vasquez maintains her encouraging spirit and therefore defends Guaspari's vision, the vision that is being propped up by the film's narrative. Even though we witness an African American character disparaged by a Latina, the Latina is a much more important character, so her warmth outweighs any unprofessionalism or even "bitchiness" the black woman represents. In this way, *Music of the Heart* offers us two visible examples (Williams and Vasquez) of likable, nonstereotypical (at least in the realm of school films) characters who are not white.

In keeping with this appearance of atypical racial politics, *Music of the Heart* places its main individual villain in the body of a white male teacher (the institutional villain would be the board of education). This teacher, Raush, is pessimistic, undemanding, rigid, uncooperative, and obstructionist—in sum, he embodies much of what one would *not* look for from a competent, responsible educator. When he lectures Guaspari early on about how limited the students are in demonstrated ability, she counters in a brief exchange:

GUASPARI: I think you're underestimating them. They can learn to play as well as any other kids.
RAUSH: Oh that I'd like to see.
GUASPARI: Well, you, will, or you would, if . . .

It is at this point that the principal is convinced enough to give her an opportunity—as a substitute, but an opportunity no less. This scene contains a lot of meaning, for it sets up Raush, a white man, as eventually the most negative force within the school, and Guaspari as a refreshing spirit who actually believes in the ability of minority kids, all under the watchful guidance of a wise and sympathetic black woman. Not a radical set of characterizations, but a notable departure from the trend established by the other school films of this era.

These aspects of *Music of the Heart* seem to call into question any simple parallels one might see between the ways in which the earlier films—even the ones from the 1990s—and this one deal with race. There are some salient distinctions between *Music of the Heart* and the others, to be sure, and any distinctiveness in this regard is important to consider. However, much like the politics of Bill Clinton during whose administration the film was released, the film abandons a more enlightened attitude toward race when it suits its purposes.

One profound example revolves around Naeem, an African American child and one of Guaspari's first students. Naeem is portrayed as a likable, even adorable character, who takes to Guaspari immediately and displays a sense of excitement about learning to play the violin. After a short while, Guaspari notices that Naeem is absent from class. She encounters him on the school playground and asks him about his absence. He informs her that he cannot be involved in the class any longer because his mother will not allow him to be. When Guaspari asks, "Why?", Naeem's mother steps into the frame and explains: "My son's got more important things to do than learn dead white men's music." The scene proceeds from there:

> *GUASPARI*: (*laughing*) They're gonna learn "Twinkle, Twinkle, Little Star."
>
> *MOTHER*: How many black classical composers can you name? How many black classical violinists do you know?
>
> *GUASPARI*: Well, that doesn't mean that's the way it should be. I mean, Naeem's just learning to play music. Maybe he's—it makes him feel good about himself. What does it matter who wrote it?
>
> *NAEEM*: Please, Ma? Can I be in the class?
>
> *MOTHER*: Look . . . I've seen this before. You white women come up here and think you can rescue all the poor, inner-city children who never asked to be *rescued* in the first place. No thank you. (*To Naeem*) C'mon.

As the mother and child walk away, Naeem takes a glance back at Guaspari. The mother then delivers the final blow: "Don't look back at her. Turn your head around." Guaspari is left standing in place, apparently flabbergasted.

The verbal exchange alone speaks for itself. But the visual component of this scene is equally relevant here. The mother, like Naeem, is African American, and her appearance is designed to say a lot about her racial identity. She is dressed in African and African-inspired clothing, with cowrie shell earrings and other vaguely ethnic jewelry, an apparently unpermed hairstyle wrapped in a printed cloth, and a shoulder bag made of mudcloth. Her attire, added to the mannerisms she exhibits, like rolling eyes and a constantly moving neck, tell the audience that this woman is "authentically" black—and not just black, but ethnically centered and proud. The film codes this character in such a profound way that it entices the audience to understand exactly what she represents. And the powerful performances from all three actors in the scene make the exchange all the more compelling. In fact, she seems to have been almost lifted from past films and grafted into this one in reconstituted form. Naeem's mother is a composite of the vocal black parent from *Lean on Me* (but less strident); and the grandmother from *Dangerous Minds* who told the teacher to "find yourself some other poor boys to save" (but more polished). She reads, in short, as the late 1990s image of the strong black female parent.

One way to interpret this scene would be to understand the mother's words as a needed corrective to the dangerous white savior mentality that permeates the narratives of *Renaissance Man* and *Dangerous Minds*, and that sometimes appears among actual white teachers of nonwhite students.[25] Such an interpretation might suggest that the makers of *Music of the Heart* learned lessons from the past films and attempted to produce a more enlightened text. But this is not the end of the story. Just a little later Guaspari sees Naeem and his mother again in the courtyard. She walks over and defends herself—meekly—against the earlier charges made by the mother. She argues that she did not enter the school with the intention of rescuing anybody but because she was a single mother and "needed the job." After that brief moment of identification (taken together, the scenes imply that Naeem's parent is also a single mother), however, Guaspari delivers a lecture of her own. "And I know you think you're protecting your son, but you're not. I mean, what if Arthur Ashe's mother said he couldn't play tennis because it's a 'white man's game'? You know, the important thing is Naeem. When he plays music he's, his whole face lights up . . . You should see that." In this provocative role reversal, it is Guaspari who walks away and the mother who is rendered speechless. The movie shows us that not only is Guaspari capable of teaching kids to play the violin, but also of teaching mothers—black mothers—how to parent.

This intricate dance of recuperation is not yet over, because not long after the second courtyard encounter Naeem's mother, full of humility, returns her son to Guaspari's classroom. She asks if Naeem can reenter the violin class, and Guaspari graciously agrees to it. The teacher is even so charitable as to offer to give Naeem private lessons at her home in order to help him make up for lost time. When the mother says that she would not be able to afford that, Guaspari indicates, with the shrug of her shoulders and a couple of softly spoken words, that the lack of compensation would not be a problem. She ends the scene, as one reviewer declares, with a "spontaneous skip of triumph" (Wloszczyna 1999). And triumph it is, because the white teacher has won over her opposition.

In one fell swoop, the film neutralizes the power of the African American woman. It is she who is remorseful, and who humbly cedes the moral high ground to Guaspari. And we might view this as an example of racial progress, in the sense that we could infer the message of sometimes needing to look beyond the narrow confines of race to see the gifts that racial Others might have to share, and more specifically that Guaspari is not like all of those other white saviors but a different kind of person altogether. However, the complication here is that the entire series of events never took place in real life. The actual Guaspari attested that no parents complained about her teaching classical music, though there seemed to be a lingering sense that introducing more ethnic music might be good (Hartocollis 1999). So, what purpose did this incident in the film serve? And at whose expense is its message conveyed? The filmmakers invested so much significance in the character of Naeem's mother and in the episode involving her that, for whatever humanizing the episode does for Guaspari's character, it is difficult to understand why this particular fictionalization needed to occur in a film whose entire function is to venerate her. This episode is only one example in the context of the entire film, but it is so racialized, and consumes so much screen time, and is so filled with meaning that it speaks volumes.

Music of the Heart includes a mixture of racial messages, some that seem to counteract stereotypes and other that powerfully substantiate them. This mixture is surely a move in the right direction toward racial justice, for challenging stereotypes necessitates presenting a range of characterizations of every group possible. Creating a narrative that makes the spectrum of human behavior available to all races is a worthy goal, and this film does that at some level. Unfortunately, the value of offering this range of characterizations is undermined by the impact of that which reinforces the already well-worn stereotypes. It is true that a Latina teacher looks beyond constructed boundaries (racial and otherwise) to assist a well-meaning white colleague, although she does so in the service of whiteness and in part at the expense of blackness. And, importantly, the film even has Principal Williams dole out a parenting lesson of her own to Guaspari. After her elder son chokes another boy in a courtyard fight, Williams tells her that her child is suffering and that she needs to deal with it. This time, the reprimand flows from the black character to the white one. This demonstrates the film's willingness, in a white–black relationship, to sometimes allow the black person to inhabit the more righteous position.

Moreover, Williams, the most supportive character in the film, is contrasted with the most villainous one, a cantankerous white educator. But the villain is such a cartoon that it is easy to root for Guaspari over him; Guaspari's more sophisticated and more provocative battles involve a black woman in the person of Naeem's mother, and we are invited see and *feel* who wins and who loses. Yes, we see white, black, and Latino characters all giving moral lessons, and we are taught that parenting advice can come from someone of any race. However, when Williams counsels Guaspari about her child the advice has nothing to do with race. When Guaspari "teaches" Naeem's mother, the lesson is entirely a racial one. In this and other ways, instead of departing from the racialized patterns already set up by previous school films, it merely pretends to challenge those earlier patterns while actually working deftly to recover a privileged racial position for whiteness.

The racial and the urban meld together in *Music of the Heart*, since by now the two have been combined into one inseparable whole in the arithmetic of the school films. The focus on the urban environment that surrounds the school and all that entails begins pretty early in the film. When Guaspari's old male friend, Brian Turner, informs her about CPE and the possibility of teaching there, he also tells her that she will have to move from where she is living. Immediately, the film gives her (and the audience) a sense of what that will mean: a taxi ride to the school allows her to witness the gritty atmosphere of East Harlem, with its loud sounds (horns, car alarms, music), abandoned buildings, and aggressive inhabitants. Viewers of past school films (and films in general) have seen this before, when a character feels the sensory impact of the urban for the first time. In this instance, the film softens the blow a bit, because what we see is not so scary right away. The streets are full of life, but not (not visibly, anyway) of crime or violence. The tone is light, in fact, with the taxi driver announcing, "Welcome to East Harlem," and Guaspari looking nervous but not quite fearful. The film develops this imagery of grittiness when Guaspari moves out of Turner's home and finds a flat near the school. The empty apartment is dirty, disheveled, and almost barren, but Guaspari informs the landlord that the place is suitable. Just as she agrees to take it, a siren roars past. The film reminds us via the sights and sounds that this is the inner city, but does not go so far as to suggest or to tell us (as do *The Principal* and *Lean on Me*) that it is a "jungle" (although Guaspari later witnesses what seems to be drug sale from her window).

This more innocuous imagery is not with us for long, though. For example, when Guaspari goes to Principal Williams to complain about a scheduling matter, the principal implies that she will address it but has more serious concerns to deal with. She pulls out a box cutter that she claims to have found on a second grade student that morning. The conditions surrounding the incident are not provided; still, the impact of the image stands. And this is only the first of a few remarkable incidents in which the students' urban lives involve or are affected by dangerous elements. Another example arises just after Guaspari accepts her new, run-down apartment. She is chastising her students for not having practiced sufficiently, and she focuses on one student in particular. After receiving her scolding, the student (who appears to be a Latina) declares that her grandmother has been mugged. When Guaspari, who at that point becomes more sensitive, asks how the grandmother is faring, the student announces that she died.

Later in the movie, after ten years or so, Guaspari and her students learn that one of their black classmates was killed accidentally in a drive-by shooting. Lastly, one of her best students (also black)—so good that Guaspari recommended her for a Juilliard audition—approaches her just weeks before the big fundraising concert. The student has to move away to a secret location, she tells Guaspari, because her mother has been suffering abuse at the hands of her father. Through the use of these incidents, the film codes the urban space as teeming with danger. What *Music of the Heart* does differently from the other school films that demonize the inner city is that it has all of its scariest events take place *outside of* the school's grounds and *offscreen*. The box cutter incident does happen in the school, but we still do not witness any consequence resulting from the presence of the knife on the campus, and it is not even clear whether the possession of the cutter was a deliberate act with malicious intent or just a benign mistake. In sum, the movie does not allow us to forget where we are, but it does spare us the extreme imagery with which the past films so willingly bombarded us.

In order to analyze further how the film portrays urban space, the issue of "realness" must come into play once again. None of the violent or near-violent incidents recounted above happened in the real life of the school during Guaspari's time there; each of them was constructed by the filmmakers. In reality, there were no knives brought to school, no grandmothers killed, and no drive-by shootings, although a student did reportedly die during the period covered by the movie (Hartocollis 1999). However, for some reason the filmmakers decided to create a tougher school for the big screen. As one might imagine, this specific aspect of the depiction of CPE was underneath many of the public complaints about the film. Founding director Meier took issue with the film's inventions, arguing that its interpretation distorted the true identity of the school. As an example, she notes that not one of the students from CPE's first six graduation classes had died violently as of 1994 (Meier 2000). The school's director at the time of the movie's release described the neighborhood as "vibrant and wonderful" as opposed to "dangerous" (*Montreal Gazette* 1999). Furthermore, *Small Wonders*, which as a documentary obligates itself to a different connection to truthfulness and accuracy, does not focus on any violence connected to the school.

The film's opening credits acknowledge that it is inspired by the documentary and based on Guaspari's life story. Thus, the fictitious incidents were expressly created for inclusion in the film. Guaspari actually expressed regret over this aspect of the movie. She claimed that the environment "that they portrayed is not like our school" (though she did serve as a consultant to the movie and claimed to be proud of the end result) (Hartocollis 1999). Elsewhere, she contended that the film's narrative is "very, very true," aside from some composite characters and except that the real CPE is "warmer and more of a community-oriented school than depicted in the movie" (Kwiatkowski 1999). It is important to note here that Guaspari has described the East Harlem community in which she lived as dangerous (at least initially), writing for a few pages in her autobiography about repeatedly listening to gunfire, having her house burglarized once, and hearing her younger son speak of having to avoid crossfire (Guaspari 1999). However, she did say that almost every calamity that involves the *students* was made up (it is unclear which ones she meant) (Hartocollis 1999). Even the studio entered the fray, perhaps only to ward off negative publicity. A senior executive at Miramax said at the time that the filmmakers probably should not have used the school's actual name in the film—which seems an odd statement since Guaspari's story would still have been a real one. The film "is not really about the actual school," the same executive argued. "The school is just the canvas. The story is about how she saved the music program . . . There is no violence that occurs in the school" (*Montreal Gazette* 1999).

We are faced once again with the question of why the filmmakers made the choices they did—and especially in altering a true story—in their depictions of an urban environment. The Miramax executive's comments tell us something, but not quite enough to answer this question definitively. As a matter of fact, the comments are instructive for what they do not communicate: they focus on the wisdom of using the name of the school, but there is no compunction expressed about the depiction itself. In *Music of the Heart* the violence occurs offscreen, the dangers lurk outside of the school's boundaries. These narrative choices may give the impression that the film is atypical because it avoids the extreme characterizations of the inner city featured in other films of the subgenre. Nonetheless, the film still relies on the association between urbanness and violence that was securely established during the Reagan era, regardless of the intensity of its depiction. Did the filmmakers not trust its audience to swallow their story without coloring the distinction between its white savior and the nonwhite, urban space in which she carries out her missionary work? Were they indulging their own fears of the inner city, or those that they imagined their audience would have? Again, the answer is unclear. It may be that the makers of *Music of the Heart* surrendered, to use Meeropol's term, to the disparagement of racial minorities and urban spaces perfected by the films that preceded it.

Music of the Heart covers a great deal of ideological ground in the two hours that it spends telling its particular story. Together with *Renaissance Man* and *Dangerous Minds*, it represents a different stage of evolution than that of the school films released under Reagan and Bush. The white protagonists in these texts are not outwardly fearful of the racial minorities for which they are responsible (even though the films tell us that they have good reason to be). Instead, they perceive the potential of their students and foster their growth, civilizing them along the way through the use of Shakespeare, Thomas, and Bach. To be sure, the portrayals in these films challenge, implicitly and explicitly, past stereotypes about nonwhites. *Music of the Heart* goes the farthest in this regard by questioning the white savior mentality and by demonstrating an unconditional belief in the ability of all races of people to succeed. It also appears to be the most liberal as it fights for the importance of the arts in education, a position that puts it at odds with the back-to-basics movement that prevailed during the height of the Reagan era. But for all of its liberal positions and counterstereotypical characterizations, the film remains carefully focused on promoting its white hero—even, like Clinton, at the expense of its nonwhite characters when necessary.

6

Learning from School Films

Throughout this study, I have concerned myself with a small subgenre of the cinema: Hollywood movies about educational institutions, otherwise called school films.[1] Specifically, I have attempted to deal with the racial representations and educational messages that these films offer to their respective viewers. The set of films under examination is small for a few reasons. First, although there is a plethora of movies whose primary setting is the campus of some sort of school, only a few of them are particularly focused on the educational process that takes place within the school. The second reason for the small number of films in this study is that I wanted to focus on mainstream, widely seen or financially successful movies. All of the movies either received critical attention and/or acclaim, or profited from large box-office earnings, or both. Third, I selected films that were released between 1980 and 2000, a period when a large number of films of this subgenre appeared on screen. As I move toward some concluding remarks, I want to discuss these criteria in a bit more detail.

Again, the first criterion for each movie is that it must focus on some act of intervention within an educational setting. Both the act of intervention and the setting are broad selection criteria, as the inclusion of *Class of 1984* and *Renaissance Man* demonstrates, but both are present in each film in the study. The presence of these elements suggests that the creators of each film intended to make more than a superficial statement about educational institutions or about education in general since they devoted an entire text to deal with this issue. Having this criterion permitted me to be able to compare the educational intervention featured in the film's narrative with the underlying messages implicitly endorsed in the dialogue, images, and other textual cues.

The choice to limit the sample to popular Hollywood movies is also crucial. Obviously, selecting films on the basis of their popularity or mainstream status is not a guarantee that they left some definite, measurable ideological imprint on each person who viewed it. Indeed, neither lucrative box-office receipts, nor numerous reviews, nor prestigious award nominations, nor frequent home video circulation is an indicator that the messages of a given movie have been wholeheartedly and uncritically embraced by a large number of viewers. However, the broad attention gained by the school films in the study does ensure that the explicit and implicit messages of the films were at least *engaged with* by a broad audience. And a large or supportive viewership for a movie suggests that, at the very least, a significant number of people found something in the film that was appealing and were affected by the film in some meaningful way. The mainstream Hollywood status of the films also serves as an indication that their racial representations likely contribute, as do those of Hollywood films in general, to the formation of "one grand, multifaceted illusion" (Guerrero 1993, 2; see also Pettit 1980).

The third criterion—that each film have a release date between 1980 and 2000—enabled me to make connections between what went on in movie theaters and in the White House during the Reagan, Bush, and Clinton administrations. These three criteria have allowed me to examine a small but important group of films in terms of their racial representations. Yet selecting the sample of films was only the first step. In order to analyze them, it was necessary to incorporate other pertinent information, such as the politics of the presidential administrations, especially regarding their educational and civil rights agendas, and the sociocultural climate of the country during the 1980s and 1990s. Equally important was a consideration of the intentions and actions of the makers of the selected films, as well as of the public responses to these films. These additional factors were all essential in drawing more nuanced, meaningful conclusions about the racial representations of the school films.

The Lessons the Films Attempt to Teach

Of interest in this study are the collection of messages that are promoted and stratagems that are utilized by the set of films. This collectivity is related to what Cortés (1992) calls the "movie curriculum." A film's "curriculum" is not simply inherent in the images, sounds, and words emerging from the screen, however; it is instead constructed through the manner in which these elements are configured by the film's writers, directors, and producers (and performers) to create the final product. This final filmic product is influenced by both the intentions and desires of the filmmakers as well as by their unconscious or unacknowledged fantasies. It is for this reason that I attempted to incorporate, where possible, information about the conditions of a film's creation. Moreover, the impact of a film does not occur in a cultural vacuum; it depends on the context in which the film appears. It is therefore essential to consider the social and political environment during the period of a film's release in order to more accurately interpret its meanings.

In light of all these factors, the question to be answered is, "What do these films attempt to 'teach,' and how do they accomplish it?" As one might expect, the school films take different approaches to convey their messages. Most of them rely on the filmic formulas of older school films such as *Goodbye, Mr. Chips*, *Blackboard Jungle*, and *To Sir, with Love*, generally resorting to variations on established themes. Much of the creativity among the films, then, rests in the settings chosen and characterizations created by the filmmakers. In the previous chapters, I have presented close, contextualized readings of the selected school films in an effort to uncover their messages with regard to race. I devoted a good deal of space to analyzing the representations of each film and to highlighting the connections among them. As a way of summarizing the detailed analyses provided earlier in the study, it is worth enumerating a few identifiable patterns among the eight films, their subtle differences notwithstanding.

The first of these patterns lies in the connections that the films make between the urban and the undesirable. Taken as a whole, all of the films in the study depict the inner city as the locus of distraction, decay, violence, and other crime. They actually go further and suggest that the best way to deal with the undesirability of the city is to exercise harsh control over its inhabitants or to avoid the city altogether. While not all the films make pointed verbal statements about the dangers of the city like the ones in *Class of 1984* (e.g., "You're not in Nebraska anymore"), the imagery included in the school movies conveys a similar message. This is arguably more significant, since Hollywood films rely on the power of the image for their impact. Whether portrayed as a "jungle" as in *Lean on Me* or, by implication, as inferior to suburban or country living as in *Dead Poets Society*, life in the city is cast in an unflattering light. Given that the terms "inner city" and "urban" are often used as stand-ins for "nonwhite," the racial significance of this tendency among the school films is apparent.

The school films of the middle and late 1990s continue to portray urbanness mainly as criminal or undesirable, but they go about it in a slightly different manner. Whereas the earlier films of the group demonize the inner city and those who inhabit it or isolate whiteness—both physically and symbolically—from urban settings, the later films such as *Dangerous Minds* and *Music of the Heart* return to the city space not only to tame the threatening urban dwellers (again, largely nonwhite), but also to restore order to their environment and refine them through canonical literature or instruction on instruments that require "discipline" (e.g., the focus on Shakespeare in *Renaissance Man*, the teaching of violin in *Music of the Heart*).

The second shared element of almost all of the films is the centrality of a white educator who is depicted as the savior of the students in the film; furthermore, in all of these films except *Dangerous Minds* and *Music of the Heart*, this white savior is a man. The films tend to focus on the affection that the students develop for their saviors, an affection that "transcends the problems encountered" and validates the saviors' respective pedagogies (Farber and Holm 1994, 159).[2] In *Lean on Me*, of course, the hero is Joe Clark, who is African American, but as I argue in chapter 3, he espouses in many ways the ideologies of Ronald Reagan and William Bennett, two white, male, American heroes. And in *Dangerous Minds*, the character of LouAnne Johnson is portrayed as being driven by the values of white men—both literally, when she relies heavily on the advice of her colleague-friend, and symbolically, as a product of the marines. In *Music of the Heart* also the central figure is a white woman who is acknowledged in grand fashion at the end of the film for "saving" the violin program. The similarity of the portraits of the hero (and especially of the white male characters in general) in the sample of school films suggests that their respective filmmakers believe that the lone, white (often male) educator is best suited to deal with the challenges residing in today's educational institutions.[3]

A third tendency observed among these films is a reliance on disparaging stereotypes of racial minorities to propel their narratives. The school films I examined share a set of character types and situations that position nonwhites—most often African Americans and Latinos—as inferior to whites or as responsible for the injustices from which they suffer. These elements show up in the earliest films in the study and resurface with greater frequency in the more recent films. For example, the finger of blame for the white teen's pregnancy in *Teachers* is pointed at the gym teacher who fathers the child, whereas the pregnant black student in *Lean on Me* is rebuffed for her misfortune and the irresponsible black teenage father is shamed; by the time we reach the seemingly more liberal *Dangerous Minds*, the requisite black teen mother is nudged out of school by the administration, only to be informed by the sympathetic white teacher that she can "choose" to return.

Racist stereotyping is also evidenced by the fact that in almost all of the films, African Americans and Latinos commit or are somehow involved in the acts that are most criminal. For instance, in *The Principal*, the main African American student plays a part in almost every criminal act that occurs on the school's grounds. In *Renaissance Man*, the smartest recruit in the remedial class—an African American man—is ultimately imprisoned for selling drugs. And an older, Latino student in *Dangerous Minds* exerts his influence over his classmates to thwart his teacher's efforts to reach them; he ends up being killed partly because of his violent dealings. Notably, before his death the Latino student attempted to follow a nonviolent path to resolving his problem, only to be ignored by his African American principal. In contrast, white students are the primary threats in two of the earlier films, *Class of 1984* and *Teachers*; however, these two films portray the white students as troubled individuals whose behavior results from their social conditions or mental health and not from their racial backgrounds. The lone exception, *Dead Poets Society*, cleverly avoids having its white students deal with violent crime and teenage pregnancy.

The stereotypes exchanged in the school films are problematic for at least two reasons. For one thing, they are featured prominently in the films but are rarely called into question. The potential harmfulness of racial stereotypes is acknowledged in one particular scene in *Lean on Me*, but the stereotyped and not the stereotypers are asked to take responsibility for eradicating them. It seems that *Renaissance Man* attempts to unpack certain stereotypes by trying to validate the experiences of the stereotyped, yet ends up falling back on other, more serious ones. In the larger scheme of things, the stereotypes are problematic because they are not usually counterbalanced by diverse or contesting representations in other school films or in Hollywood in general. *Music of the Heart* is atypical in the degree to which it offers a variety of representations of characters of multiple races. "Instead of efforts to construct a truly universal system of communication that builds egalitarian understandings between diverse groups and cultures," as Guerrero (1993, 2) suggests should be happening, audiences contend with the products from an industry that remains narrowly focused and inegalitarian.

One school film that diverges ideologically from the others in this study regarding race—and, hence, has not been analyzed here—is *Stand and Deliver*. Although the movie's narrative is far from radical and was even applauded by the Reagan administration, it does not resort to the same racial politics of its counterparts. This is perhaps because the film's origins lie outside of the Hollywood mainstream, in that it was initially intended to be released as part of PBS's "American Playhouse" series.[4] Another possible explanation is that its writer-producer (Tom Musca) and Cuban American writer-director (Ramon Menendez) are more sensitive about the consequences of racial representations in films because of their own ethnic identifications, and this increased sensitivity is thus reflected in their movie.

Stand and Deliver depicts an all-Latino (presumably Chicano) public high school in East Los Angeles, with its attendant delinquency, graffiti, and low academic achievement—but it does so without focusing narrowly on the criminal element or without employing the racially biased tactics of the other school films. Apparently, this was the intention of the filmmakers, according to a statement by the writer-director: " 'Maybe the world at large will see the film and realize something about the Latino experience, and see these people as human beings just like them' " (Kroll 1988, 62).[5] This sentiment certainly differs from those expressed by the makers of *Lean on Me* and *Dangerous Minds*. Does this mean that white filmmakers are not capable of producing counterhegemonic racial representations? No, but it does mean that they have to make a special effort to pursue progressive racial politics in their films and that they must be held accountable for the representations they do produce—as do nonwhite filmmakers.

All in all, the three aforementioned patterns work in conjunction with each other to create two distinct filmic spaces: one for nonwhites, which is coded as urban and rife with poverty, violence, crime, teenage pregnancy, and academic inferiority; and another space, perhaps resembling

Nebraska (as alluded to in *Class of 1984*) or a rural East Coast town (as in *Dead Poets Society*), where order and serenity reign, definitely inhabited by whites. The school movies under examination follow these patterns—in a roughly chronological fashion and closely matching the corresponding political climate—in a move toward the reclamation of the privileged racial position of whiteness, all the while perpetuating the ideologies of the Reagan era.

It is worth reiterating that the ideologies of the school films in question resonate strongly with the dominant ideologies and discourses that have circulated since Reagan's entry into the White House. One can find counterparts within the Reagan legacy to the three patterns that I have identified in this body of films. Some of the films are more closely or directly linked to Reagan's leadership than others, but all of them are connected to the Reagan era in some significant way. This correspondence between the dominant political agendas and the mainstream cultural texts prevalent during the time of President Reagan and thereafter underscores the significance of the representations in the school films: it highlights the manner in which representations both reflect and influence the ideologies of society, and the ways representations function to support certain interests and to repress others.

That my analyses of the films in this study either suggest or outrightly state that there is a privileging of white people in each film or in general is no accident. This should not be surprising to any student of racial representations in Hollywood films. To identify this underlying white supremacy, however, is not to make the simple assertion that "these films are racist." I concur with Stuart Hall that "it would be wrong and misleading to see the media [cinema in this case] as uniformly and conspiratorially harnessed to a single, racist conception of the world" (1981, 35). I have thus attempted to illustrate the powerful, complex, and sometimes contradictory racial politics of the body of school films. The import of this study, then, lies in its exploration of the degree to which this racial supremacy is asserted, and of the various ways it is asserted in specific movies and in the main. My intention was to examine the representation of race in a sample of school films, and my hope is that the reader will at some level find the results persuasive.

Obviously, movies are not the only cultural texts in which these sorts of racial representations occur; oppressive racial politics can be observed in, say, literature, television, or even advertising, as the studies of Morrison (1992), Gray (1995), and Green (1993) demonstrate. Yet the increasing availability of movies via home video rentals, cable showings, and television broadcasts leaves the public all the more exposed to their messages. This leads to a crucial concern over how my study of the racial representations in school films might be beneficial to the very public being exposed to them.

Learning from and Teaching about School Films

As I argued earlier in this chapter, these school films promote certain messages through the images, character portrayals, musical selections, and other elements of which they are comprised. In this way, the films make statements about race (among other things) that are far from coincidental or innocent. Yet most of us recognize that what is "learned" from a particular film depends on the specific set of experiences, sensitivities, and goals that a given spectator brings to the viewing situation. "Accordingly," as Moore (1991) writes, "teachers need to be as involved with how things are learned as with what is being offered for learning" (187–188). Hence, deciphering what is taught by and learned from these films and from films in general is a complex task and not something that can be taken lightly or performed hastily. If the films in this study do indeed offer the kinds of problematic racial representations that my analysis indicates, then there is much at stake in terms of how we use them educationally.

What and how can educators—meaning all of us involved in the exchange of knowledge—teach about these films, apart from what the films themselves try to teach? There are several ways in which to approach the task of teaching about the films in the study. First of all, the school films might be exploited pedagogically by incorporating the study of them into the curricula of current and future educators. Of course, films and other audiovisual media have been used in educational institutions for years, as evidenced in part by the abundance of media labs and audiovisual storehouses in elementary and secondary schools, colleges, and universities. Often, however, films are seen as mere replacements for overworked, fatigued, or absent instructors. At the same time, the images, representations, and information the films offer frequently go unexamined or unproblematized. Or if the films are actually discussed and critiqued, they may ultimately be perceived as supplemental or even inferior to printed texts. Instead, these school films can be used productively by present and future educators at two levels: through an analysis of the explicit statements about education that the films proffer; and, perhaps using this study as a model, through an examination of the implicit messages of the films by considering them within a broader cultural context.

Let us take *Dangerous Minds* as an example. On one level, the story of LouAnne Johnson's and her students' accomplishments during her tenure at Carlmont High is fascinating and instructive in terms of its portrayal of meaningful education in the face of adversity. It points to the productive possibilities of public schooling, and could also be viewed as a useful model of interracial cooperation within the classroom. The story also illustrates the benefit of love and caring, two oft-forgotten elements of rewarding teaching (Noddings 2005). As a matter of fact, I have encountered a plethora of accounts of educators who have been inspired by Johnson's experiences.[6] So I certainly do not want to ignore her impact on her students; indeed, I congratulate her attempts to encourage her students' investment in schooling while simultaneously contesting the policies that perpetuated the conditions that impeded her students' progress. However, I am less supportive of some of the fictional Johnson's practices in the film—like her choice of poetry for her class, for instance. Thus, although perceptions will differ, these and other aspects of her teaching can be evaluated by practicing educators (for example) through the lenses of their own lived experience.

Fortunately, some useful models of this kind of pedagogical use of school films already exist. James Trier, for instance, has documented his employment of a large body of films in the preparation of teachers. He has made use of the movies in a variety of ways, such as by exploring specific concepts through the use of various films (2002), and by creating compilations of movie excerpts that are grouped along thematic lines and then using the compilations to promote critical thinking about those themes (2003). These examples underscore the benefits of analyzing filmic texts and of incorporating these analyses into our curricula.

At the second level, the analysis would go beyond a critique of Johnson's educational practices or ideologies to examine how Johnson's teaching is depicted in *Dangerous Minds* and how this depiction serves to promote the agenda of the filmmakers. By placing the film within the context of other school films, other Hollywood films, and the country's sociopolitical climate at the time of the film's release, an instructor and his or her students could study the film's function as a cultural text. They could then understand how the filmmakers altered Johnson's story and, as a result, created a new narrative embedded with new ideologies. Examining these ideologies is a necessary step in comprehending how they undergird the larger racial politics of the creators of the film. Even if the students are in disagreement with the conclusions of this study or of the instructor, they may still benefit from the process of critiquing the film and teasing out its possible implications. Once the students become educators themselves, they will be able to carry this process of demystification into their own classrooms, whatever form these classrooms may take.

Second, this study could be of use to observers of cinematic production—that is, potential filmmakers, instructors, and practitioners. I believe that my work emphasizes the importance of other aspects of filmmaking besides the technical concerns in evaluating a film. It underscores the interrelatedness of technical as well as contextual and structural concerns, content development, audience response, and other factors that comprise the filmmaking and viewing experiences. Undoubtedly, many students and teachers involved in film production are aware of the relationship between all of these elements. One may lose sight of this, however, as the subject matter is broken down for ease of instruction (Buckingham 1991). An examination of their relationship such as this study can help to demonstrate how the elements are tied together and to reinforce the fact of their connectedness.

My analysis can also help future and current filmmakers to appreciate the fact that they have obligations, other than personal or economic ones, to a larger society for the products they create. They should understand that they must accept responsibility for their creations and recognize the possible implications of the images they present on screen. This does not mean that filmmakers should feel inhibited about exploring certain issues—race, for instance—for fear of rebuke from the "ideological police," but they should realize that their work is not released into a special space where history and social relations disappear once audiences enter a movie theater. This would help filmmakers to be creative and socially responsible at the same time. It should go without saying that this prescription also applies to producers of other media forms.

Third, I believe my study has implications for all spectators. I am in agreement with Raymond Williams's contention that the study of film should not simply be about "training more appreciative consumers" (1979, 13); that job, it seems, is being performed sufficiently by advertisers. What I set out to do was to expose the messages of a particular set of films: what racial images the films present, what activities were involved in their production, and how they are linked to what was going on in the world at the time of their dissemination. In other words, I attempted to uncover the fact that these films are cultural artifacts, economic products, and political statements—and not just innocent texts. It is my hope that viewers will learn something valuable from the study about the intersection of culture and politics, and that this will influence the way they see films in the future. Equipped with this knowledge, spectators may be encouraged to demand more responsible texts, or even to create these texts themselves.

At the very least, spectators may be aided in creating a space for themselves where the films they view can be interpreted in a *different* way than the filmmakers may have intended. Understanding the constructed nature of films helps us to avoid letting our minds and bodies be constrained by the world depicted on the screen. This understanding permits us to think and feel in ways other than those encouraged in films, and reminds us that we are empowered agents who may be influenced, but not ruled, by the power of representations.

However, my goal is not to tell spectators what to "get" out of these films. I realize that the cinema is by design a form of entertainment, and people will be entertained for countless reasons. What one takes away from a film depends heavily on one's purposes for viewing it. Nevertheless, what appears on screen has important consequences beyond its entertainment value. Our actions and decisions—from social interactions to policymaking—are based on our attitudes and beliefs, and filmic representations are capable of shaping those beliefs, the degree to which this occurs being debatable. Again, what I see as the primary aim of my work is to point to the messages imbedded in particular Hollywood school films—in short, what these films *aim* to teach. The main point is that the final product that is projected on the silver screen is not accidental, but the result of hours of manipulation and construction, and, as such, reflects in some way the fantasies of its creators and of its audience. In the sample of films I examine, these fantasies involve the perpetuation of the racial politics of the Reagan–Bush legacy.

Notes

1 Introduction

1. See Thompson (1988) for a neoformalist position on the function of art.
2. This understanding of ideology appears to be close to Balkin's (1998) "*ambivalent* conception of cultural software" (Balkin dissolves the study of the ideological realm into the study of "cultural software").
3. *Goodbye, Mr. Chips* was actually a British production, but it received critical acclaim and several Oscar nominations (and one award) in the United States.
4. I realize that some define popular culture as that which originates among the masses or among subordinated cultures (also sometimes referred to as folk culture) (see Fiske 1989, for example) as opposed to that which is created by industries for commercial purposes (such as Hollywood school films, for example). However, I identify the school films as being part of popular culture because, generally speaking, large numbers of the population have access to these texts and choose to engage with them and to make use of them regardless of their origins.

2 Reagan, 'Rithmetic, and Race

1. In a speech delivered before the conservative Heritage Foundation, then-Secretary of Education Bennett spent his time discussing economics and foreign policy, patriotism, the loss of strong families, and drug abuse. At one point, Bennett made a reference to the deaths of athletes Len Bias and Don Rogers without naming them, subtly linking drug use to black athletes. Theirs were certainly not the only deaths resulting from drug abuse: if Bennett was short on examples he could have mentioned any of the 147 other deaths that

occurred during the nine days in June that included the deaths of Bias and Rogers, most of which went unreported by the press. The majority of these victims were not athletes, musicians, or inner-city youths; a composite portrait instead depicts a 33-year-old white male with a steady job (Emmons et al. 1987).

2. The term is from Meeropol (1998, 52); see page 35 for an example.
3. It was not until June 21, 2005, that Edgar Ray Killen was convicted on three counts of manslaughter and received the maximum sentence of 60 years—20 for each count—without the possibility of parole until 2025 (Jubera and Bentley 2005).
4. I am using the expression "civil rights" to refer to *racial* equality that is protected by the government and/or guaranteed under the Constitution. I understand that the expression has been used both more broadly and more narrowly than it is here.
5. William French Smith, who departed the office in 1984, was quickly replaced by Edwin Meese, who was no less vigilant in undoing past civil rights work (Simmons 1984; Shull 1989).
6. While there is disagreement over whether busing is in the best interest of racial minorities, Reynolds's actions were *definitely* not in their best interest.
7. Days was the Assistant Attorney General for Civil Rights under President Carter from 1977 to 1980.
8. Congress (and not Reagan) increased funding for Chapter 1 in the 1988 School Improvement Act (Hylton 1988).
9. The later reactions in 1989 to the particular works of Andrés Serrano and Robert Mapplethorpe were even more vehement and consequential.
10. A quote from Nancy Reagan may be relevant here. At a reception of Reagan supporters in Chicago in February of 1980, Mrs. Reagan wished aloud (via speakerphone) to her absent husband that he could be there to "see all these beautiful white people." She reportedly hesitated, then corrected herself: "Beautiful black and white people" (quoted in Dugger 1983, 202). Perhaps this is an example of how the beautiful was the familiar.

3 Controlling the Savage Enemy: Punishing the Other and Criminalizing Urban Space

1. For more on Reagan's views of his role as president, see Orman (1987), Rogin (1987), and Dugger (1983).
2. The administration's responses were not always consistent, of course. One particularly powerful example was highlighted in 1996 by reporter Gary Webb in a three-part series for the *San Jose Mercury*

News. His reporting alleged that funders of the CIA-supported Nicaraguan Contras included traffickers of crack cocaine into South Central Los Angeles and other inner-city neighborhoods in the early 1980s (which, as a result, helped to fuel the crack epidemic in African American urban communities). Despite a few exaggerations and minor flaws, Webb's series was solid, well-sourced, new work that had built upon earlier reportage by Robert Parry and others. Unfortunately, the *Washington Post, New York Times*, and *Los Angeles Times* produced dishonest attacks on Webb's series and the *Mercury News* consequently distanced itself from the reporting.Webb's partial vindication came in 1999 when, "[i]n secret congressional testimony, senior CIA officials admitted that the spy agency turned a blind eye to evidence of cocaine trafficking by U.S.-backed Nicaraguan contra rebels in the 1980s and generally did not treat drug smuggling through Central America as a high priority during the Reagan administration" (Parry 2000).

3. O'Reilly (1995) covers well the timeline of the development of the William "Willie" Horton ad and the Bush campaign's use of it.
4. See Siegel (1997) for a discussion of why Los Angeles, perhaps unreflectively, has been connected with urban decline. He writes, more specifically, that "South Central has displaced the South Bronx as the country's favorite symbol of urban failure" (115).
5. *Class of 1984* could even be interpreted as a spoof of school films of old—of *Up the Down Staircase* (1967) in particular. Yet, even in its parodying elements, it makes certain messages about race, violence, and schooling that should not be ignored.
6. The possible allusion to Dorothy's widely familiar phrase, "Toto, I have a feeling we're not in Kansas anymore" (*The Wizard of Oz*, (1939)) is not lost here. Those familiar with the film know what perils awaited Dorothy once she landed in Oz.
7. In his review of *Class of 1984*, Paul Taylor (1983) also points out "the racial angle to the school gangs' power struggle."
8. See also Farber and Holm (1994).
9. For excellent analyses of the perceived pathological value system of urban blacks and how this perception shapes policymaking, consider all the essays in Lawson (1992).
10. *Lean on Me* was likely prompted by *Stand and Deliver* (1988) as well, another Warner Brothers release concerned with underprivileged nonwhite students, test scores, and uplift. *Stand and Deliver* is briefly discussed in the final chapter.
11. David Kirp (1989) echoes this contention when he writes, "Joe Clark and late 1980s America look to be made for each other."
12. In a 1986 speech delivered to the conservative Heritage Foundation, Bennett spoke as much or more about drug usage

than about any other social phenomenon. In a similar vein, while at Eastside, Clark declared that he spends so much time policing against vandalism, sex rings, and drug use that he "can't get to instruction. It's a shame, but it's an unimportant issue compared to everything else that's going on" (Gellene 1989).

13. Stewart (1990) addresses the "absence of sophisticated models of internal family dynamics" (23) in the extremely useful but partially flawed analytical approaches of W.E.B. Du Bois and E. Franklin Frazier. Most disturbingly, the weaknesses of Frazier's approach are what undergird the bleak and wrongheaded "Moynihan Report" (U.S. Department of Labor 1965) that has become the "cornerstone of the neoconservative opposition to liberal-sponsored government intervention on behalf of black urban poor people" (Lott 1992, 71). See also Brown (1999), especially pages 367–368.
14. O'Brien's absolution of the film's "aesthetic sins" recalls, inversely, Clyde Taylor's wise critique of film scholars who celebrate the aesthetic while ignoring the ideological, particularly with regard to D. W. Griffith's *The Birth of a Nation*. Here, as with *The Birth of a Nation*, "the aesthetic operates simultaneously and harmoniously with a kindred ideology" (Taylor 1991, 29).

4 Isolating and Protecting Whiteness

1. See also Chrisman (1991–1992).
2. In spite of the fact that the filmmakers have taken particular lines of poetry out of context and thus reinterpreted them (see, e.g., Carton 1989), it is reasonably clear what they mean for the lines to represent.
3. The spelling of Dalton's new name—whether Nuwanda or Newanda or Nuanda—is unclear in the film, but the sound of it speaks volumes.
4. See Lindsay (1963) for the complete text of the poem.
5. The relative absence of women functions similarly for the film.

5 Bringing the West to the Rest: The Transmission of Civilizing Knowledge

1. See Ladd (1993) for example.
2. For Reagan on welfare, see Dugger (1983), especially pages 285–342.

3. For a detailed account of the "worst thing Bill Clinton has done"—that is, the folly of the 1996 welfare law—see Edelman (1997).
4. Meeropol (1998) identifies the "stringent anti-inflation policy" of the Federal Reserve System launched in 1979 under Paul Volcker's leadership as being as much of a factor in the Reagan Revolution as were Reagan's economic policies, hence the term "Volcker–Reagan" (7).
5. Maraniss's account (1995) of Clinton's upbringing is informative on this issue.
6. It is obviously important to understand that these could be interpreted as votes for the Democratic Party as well.
7. Rock's comments were made in more than one forum; here, they are taken from his appearance on the "Today" show (NBC, August 19, 1998).
8. O'Reilly (1995) offers a wonderful account of the racial politics of the 42 American presidents (before George W. Bush, of course).
9. For a corrective to the overblown claims of crisis, see Berliner and Biddle (1995).
10. Reich's (1999) introduction to his ponderation on Third Way politics is relevant here: "Is the Third Way a new public philosophy likely to shape capitalism in a postcommunist twenty-first century? Or is it, as some from both ends of the political spectrum suspect, little more than a watered-down version of Reaganism–Thatcherism: less a new movement than a pragmatic, if not cynical, means of keeping liberals mollified while continuing the long-term shift rightward—a global version of Dick Morris's 'triangulation'?" (46).
11. See also Gresson (1996) for his examination of the "recovery" at work in *Forrest Gump* (1994).
12. *To Sir, with Love* was, like *Goodbye, Mr. Chips*, originally a British production; it was subsequently released in the United States and became quite successful and popular here.
13. For a very perceptive reading of how the film works in the service of the army, see the review of the film by Tunney (1994).
14. All box-office figures are compiled from *Variety*.
15. All music airplay and sales figures are compiled from *Billboard*.
16. Credit for suggesting this last example goes to David Sterritt (1995).
17. See Keroes (1999) for an exploration of the dynamics of gender (and authority and desire) in fiction and film representations of teachers.
18. CNN, "Talk Back Live," October 24, 1995.
19. Given the film's remarkable box-office draw, one would hope that the filmmakers donated some of the proceeds to Carlmont High School's Academy, particularly since the $100,000-per-year

program faced a crippling blow from federal and state budget cuts (reported in Dezell 1995).

20. The real-life teacher sometimes used her married name, Roberta Guaspari-Tzavaras, and the film alters this name by having the principal call her Guaspari-Demetras. However, the last name by which she is known in the majority of the press accounts and in most of the movie is Guaspari. For the sake of consistency, both the character and the real-life person will be referred to here as Guaspari.
21. Another was *Mr. Holland's Opus* (1996), a film that is not analyzed here. Its racial politics are notably different from the other films from the 1990s.
22. Miramax, of course, was until recently an independently run unit of the Walt Disney Company. The Weinstein brothers founded the company in 1979 as an independent distribution and filmmaking company; they sold it to Disney in 1993. The Weinsteins and Disney officially parted ways in March of 2005 (Mohr 2005).
23. Examples include *The Crying Game* (1992), *Pulp Fiction* (1994), *The English Patient* (1996), *Good Will Hunting* (1997), and *Shakespeare in Love* (1998).
24. According to Guaspari's (1999) autobiographical account, the student population of CPE was mostly African American and Latino with a small but growing Asian American population and a white population of 10 to 15 percent.
25. See McIntyre (1997) for example.

6 Learning from School Films

1. Examining the entire subgenre is obviously a worthwhile project as well. Dalton's work (2004), for example, surveys a much larger portion of these films.
2. Farber and Holm (1994) also note the presence of this "educator-hero," referring to this group of male saviors as "a brotherhood of heroes" (due to the fact that the study was done before *Dangerous Minds* and *Music of the Heart* were released). However, their study diverges from mine in that they focus on the educator-hero and not specifically on race and therefore use different criteria for selecting their sample of films.
3. Because of this shared element and others, one might argue that these films also embody the "American monomyth" that Jewett and Lawrence (1988) identify so cogently in popular culture.
4. "*Stand and Deliver*: PBS Night at Movies," *Broadcasting*, March 13, 1989, 41.

5. See Noriega (1992) for an analysis of the film reviews of *Stand and Deliver* and the different ways in which mainstream, alternative, and Hispanic press outlets discussed the film's racial representations.
6. See, for example, Saillant (1995), Glass (1995), Hering (1995), Hunt (1995), and John (1995). Some of these encounters have also been conversations with teachers with whom I am acquainted.

Bibliography

Balkin, J. M. 1998. *Cultural Software: A Theory of Ideology*. New Haven, CT and London: Yale University Press.

Battiata, Mary. 1985. "NEA's Pornography Ruckus." *Washington Post*, sec. C, September 12.

Beezer, Anne, Jean Grimshaw, and Martin Barker. 1986. "Methods for Cultural Studies Students." In *Introduction to Contemporary Cultural Studies*. Edited by David Punter, 95–115. New York: Longman.

Bell, Terrel H. 1986. "Education Policy Development in the Reagan Administration." *Phi Delta Kappan* 67: 487–493.

Bennett, William. 1988. *Our Children and Our Country: Improving America's Schools and Affirming the Common Culture*. New York: Simon and Schuster.

Bensman, David. 2000. *Central Park East and Its Graduates: "Learning by Heart."* New York: Teachers College Press.

Berg, Charles Ramírez. 1990. "Stereotyping in Films in General and of the Hispanic in Particular." *Howard Journal of Communications* 2: 286–300.

Berliner, David C., and Bruce J. Biddle. 1995. *The Manufactured Crisis: Myths, Fraud, and the Attack on America's Public Schools*. Reading, MA: Addison-Wesley.

Biskind, Peter. 2004. *Down and Dirty Pictures: Miramax, Sundance, and the Rise of Independent Film*. New York: Simon and Schuster.

Bloom, Allan. 1987. *The Closing of the American Mind: How Higher Education Has Failed Democracy and Impoverished the Souls of Today's Students*. New York: Simon and Schuster.

Blumenthal, Sidney. 1988. "Reaganism and the Neokitsch Aesthetic." In *The Reagan Legacy*. Edited by Sidney Blumenthal, and Thomas Byrne Edsall, 251–294. New York: Pantheon Books.

Bogle, Donald. 1973. *Toms, Coons, Mulattoes, Mammies, and Bucks: An Interpretive History of Blacks in American Films*. New York: Viking.

Bordwell, David, Janet Staiger, and Kristin Thompson. 1985. *The Classical Hollywood Cinema: Film Style and Mode of Production to 1960*. New York: Columbia University Press.

Britt, Donna. 1995. "Finding Truth among Hollywood Lies." *Newsday*, August 13.

Brown, Michael K. 1999. *Race, Money, and the American Welfare State*. Ithaca, NY: Cornell University Press.

Buckingham, David. 1991. "Teaching about the Media." In *The Media Studies Book*. Edited by David Lusted, 12–35. London and New York: Routledge.

Bush, George H. W. 1993. Remarks to Community Leaders in Los Angeles, May 8, 1992. *Public Papers of the Presidents of the United States: George H. W. Bush, 1992–1993*. Washington: U.S. Government Printing Office.

Canby, Vincent. 1989. "Subversive Verse." Review of *Dead Poets Society. New York Times*, sec. C, June 2.

Cannon, Lou. 1980. "Reagan Campaigning from County Fair to Urban League." *Washington Post*, final edition, sec. A, August 4.

Carr, Jay. 1989. "Rocky Goes to School." Review of *Lean on Me. Boston Globe*, March 3.

Carton, Evan. 1989. "Better Dead than Read: The Society of Poets." Review of *Dead Poets Society. Tikkun*, November–December, 64–67.

Cavazos, Lauro. 1990. "More Parenting, Not More Money." *New Perspectives Quarterly* 7 (Fall): 6–8.

———. 2002. "Emphasizing Performance Goals and High-Quality Education for All Students." *Phi Delta Kappan* 83: 690–697.

Chapelle, Tony. 1989. "The Benevolent Dictatorship of Joe Clark." *Black Collegian*, March–April, 68–70, 125.

Chrisman, Robert, ed. 1991–1992. "The Clarence Thomas Confirmation." Special Issue, *Black Scholar* 22 (1–2).

Christopherson, Susan. 1994. "The Fortress City: Privatized Spaces, Consumer Citizenship." In *Post-Fordism: A Reader*. Edited by Ash Amin, 409–427. Cambridge: Blackwell.

Clark, Ginger. 1996. "Beyond the Great Divide: Hollywood and Race Relations in the 1950s." *Howard Journal of Communications* 7: 373–382.

Clinton, William J. 1991. Keynote Address to the Democratic Leadership Council's Cleveland Convention, May 6. http://www.dlc.org/ndol_ci.cfm?kaid=86&subid=194&contentid=3166.

———. 1996. *Between Hope and History: Meeting America's Challenges for the 21st Century*. New York: Times Books.

Collins, Ava. 1994. "Intellectuals, Power, and Quality Television." In *Between Borders: Pedagogy and the Politics of Cultural Studies*.

Edited by Henry A. Giroux, and Peter McLaren, 56–73. London and New York: Routledge.

Collins, Mark. 1989. "Make-Believe in *Dead Poets Society.*" *English Journal* 78 (8): 74–75.

Cortés, Carlos E. 1992. "Who is Maria? Who is Juan? Dilemmas of Analyzing the Chicano Image in U.S. Feature Films." In *Chicanos and Film: Essays on Chicano Representation and Resistance*. Edited by Chon A. Noriega, 83–104. New York: Garland.

———. 2004. "Knowledge Construction and Popular Culture: The Media as Multicultural Educator." In *Handbook of Research on Multicultural Education*. 2nd ed. Edited by James A. Banks, and Cherry A. McGee Banks, 211–227. New York: Jossey-Bass.

Crenshaw, Kimberlé Williams. 1988. "Race, Reform, and Retrenchment: Transformation and Legitimation in Antidiscrimination Law." *Harvard Law Review* 101: 1331–1387.

Dalton, Mary M. 2004. *The Hollywood Curriculum: Teachers in the Movies*. Rev. ed. New York: Peter Lang.

Days, Drew S., III. 1984. "Turning Back the Clock: The Reagan Administration and Civil Rights." *Harvard Civil Rights-Civil Liberties Law Review* 19: 309–347.

Dawes, Amy. 1989. Review of *Dead Poets Society. Variety*, May 31, 26.

Denzin, Norman K., and Yvonna S. Lincoln. 2000. "Introduction: The Discipline and Practice of Qualitative Research." In *The Handbook of Qualitative Research*. 2nd ed. Edited by Norman K. Denzin, and Yvonna S. Lincoln, 1–29. Thousand Oaks, CA: Sage.

Dezell, Maureen. 1995. "Names and Faces." *Boston Globe*, August 22.

Doten, Patti. 1989. "For Madison Park Head, Teamwork Comes First." *Boston Globe*, March 3.

D'Souza, Dinesh. 1995. *The End of Racism: Principles for a Multiracial Society*. New York: Free Press.

Duffy, Michael, and Dan Goodgame. 1992. *Marching in Place: The Status Quo Presidency of George Bush*. New York: Simon and Schuster.

Dugger, Ronnie. 1983. *On Reagan: The Man and His Presidency*. New York: McGraw-Hill.

Dyson, Michael Eric. 1992. "Between Apocalypse and Redemption: John Singleton's *Boyz N the Hood*." *Cultural Critique* 21 (Spring): 121–141.

Edelman, Peter. 1997. "The Worst Thing Bill Clinton Has Done." *Atlantic Monthly*, March, 43–58.

Edelman, Rob. 1985. Review of *Teachers. Cineaste* 14 (2): 11–12.

Edsall, Thomas Byrne. 1988. "The Reagan Legacy." In *The Reagan Legacy*. Edited by Sidney Blumenthal, and Thomas Byrne Edsall, 3–49. New York: Pantheon Books.

Edsall, Thomas Byrne, with Mary D. Edsall. 1991. *Chain Reaction: The Impact of Race, Rights, and Taxes on American Politics*. New York and London: Norton.

Emmons, Karen, Linda Gomez, and Edward Barnes. 1987. "Nine Days in June." *Life*, January, 83–85.

Farber, Paul, and Gunilla Holm. 1994. "A Brotherhood of Heroes: The Charismatic Educator in Recent American Movies." In *Schooling in the Light of Popular Culture*. Edited by Paul Farber, Eugene F. Provenzo, Jr., and Gunilla Holm, 153–172. Albany: State University of New York Press.

Farley, Ellen. 1989. "Poetry in Motion Pictures? In Robin Williams's New Film, the Dialogue Could Be Verse." *Washington Post*, sec. G, June 11.

Finke, Nikki. 1989. "*Lean on Me*: A Modern Myth? Principal and Actor See Things Eye to Eye." *Los Angeles Times*, March 3.

Fiske, John. 1989. *Understanding Popular Culture*. Repr., Boston: Unwin Hyman, 1991.

Frankenberg, Ruth. 1993. *White Women, Race Matters: The Social Construction of Whiteness*. Minneapolis: University of Minnesota Press.

Franklin, John Hope. 1998. "Talking, not Shouting, about Race." Editorial. *New York Times*, sec. A, June 13.

Friedrich, Otto, and Ezra Bowen. 1988. "Getting Tough." *Time*, February 1, 52–58.

Gellene, Denise. 1989. "*Lean on Me*: A Modern Myth? Taking Artistic Liberty with the Real Joe Clark." *Los Angeles Times*, March 3.

Gilens, Martin. 1999. *Why Americans Hate Welfare: Race, Media, and the Politics of Antipoverty Policy*. Chicago: University of Chicago Press.

Giroux, Henry A. 1993. *Living Dangerously: Multiculturalism and the Politics of Difference*. New York: Peter Lang.

———. 1994. *Disturbing Pleasures: Learning Popular Culture*. New York: Routledge.

———. 1995. "Racism and the Aesthetic of Hyper-Real Violence: *Pulp Fiction* and Other Visual Tragedies." *Social Identities* 1: 333–354.

Giroux, Henry A., Roger I. Simon, and contributors. 1989. *Popular Culture, Schooling, and Everyday Life*. New York, Westport, CT, and London: Bergin & Garvey.

Glasgow, Douglas G., and Betty Collier-Watson. 1988. Letter to the Editor. *Policy Review* 46: 92–93.

Glass, Jon. 1995. "*Dangerous Minds* Inspires Teacher." *Norfolk Virginian-Pilot*, sec. B, September 2.

Gray, Herman. 1995. *Watching Race: Television and the Struggle for "Blackness."* Minneapolis: University of Minnesota Press.

Green, Michael K. 1993. "Images of Native Americans in Advertising: Some Moral Issues." *Journal of Business Ethics* 12: 323–330.

Green, Tom. 1995. "*Minds* Makes Most of Music, Star Power." *USA Today*, sec. D, August 15.

Gresson, Aaron David, III. 1995. *The Recovery of Race in America*. Minneapolis: University of Minnesota Press.

———. 1996. "Postmodern America and the Multicultural Crisis: Reading *Forrest Gump* as the 'Call Back to Whiteness.' " *Taboo* 1 (Fall): 11–34.

Greider, William. 1992. "F Troop: The GOP Raids Public Schools." *Rolling Stone*, August 20, 25–26, 62.

Guaspari, Roberta. 1999. *Music of the Heart*. With Larkin Warren. New York: Hyperion.

Guerrero, Ed. 1993. *Framing Blackness: The African American Image in Film*. Philadelphia: Temple University Press.

Guinier, Lani. 1989. "Keeping the Faith: Black Voters in the Post-Reagan Era." *Harvard Civil Rights-Civil Liberties Law Review* 24: 393–435.

Hall, Stuart. 1981. "The Whites of Their Eyes: Racist Ideologies and the Media." In *Silver Linings: Some Strategies for the Eighties*. Edited by George Bridges, and Rosalind Brunt, 28–52. London: Lawrence and Wishart.

———. 1992. "Race, Culture, and Communications: Looking Backward and Forward at Cultural Studies." *Rethinking Marxism* 5 (1): 10–18.

Harrison, Bennett, and Lucy Gorham. 1992. "Growing Inequality in Black Wages in the 1980s and the Emergence of an African-American Middle Class." *Journal of Policy Analysis and Management* 11: 235–253.

Hartocollis, Anemona. 1999. "To Celebrate Music Teacher, Film Toughened Up School," *New York Times*, late edition-final, sec. B, October 27.

Hering, Anaklara. 1995. "*Dangerous Minds* Hits Hard." *Fort Lauderdale Sun Sentinel*, September 1.

Herrnstein, Richard J., and Charles Murray. 1994. *The Bell Curve: Intelligence and Class Structure in American Life*. New York: Free Press.

Hirsch, E. D., Jr. 1987. *Cultural Literacy: What Every American Needs to Know*. Repr., New York: Houghton Mifflin, 1988.

Holden, Stephen. 1989. "Film View: Today's Hits Yearn for Old Times." *New York Times*, sec. 2, August 13.

Horne, Gerald. 1994. "Ensuring a True Multiculturalism." In *State of the Union 1994: The Clinton Administration and the Nation in*

Profile. Edited by Richard Caplan, and John Feffer, 184–199. Boulder, CO: Westview.

Howe, Desson. 1989. "*Dead Poets Society*: Mr. Williams's English Lit." Review of *Dead Poets Society. Washington Post*, Weekend section, June 9.

———. 1994. "It's a Job, No Adventure." Review of *Renaissance Man. Washington Post*, Weekend section, June 3.

Hunt, Susan. 1995. "*Dangerous Minds*' Abstract Ideas Tinged with Reality." *Phoenix Gazette*, sec. B, September 20.

Hylton, Richard D. 1988. "Will the Class of 2000 Make the Grade?" *Black Enterprise*, October, 130–139.

Jackson, Kenneth T. 1985. *Crabgrass Frontier: The Suburbanization of the United States*. New York: Oxford University Press.

Jeffords, Susan. 1994. *Hard Bodies: Hollywood Masculinity in the Reagan Era*. New Brunswick, NJ: Rutgers University Press.

Jewett, Robert, and John Shelton, Lawrence. 1988. *The American Monomyth*. 2nd ed. Lanham, MD: University Press of America.

John, Joanna. 1995. "Counterpunch: *Dangerous* Tells It Just Like It Is." *Los Angeles Times*, sec. F, September 18.

Johnson, LouAnne. 1992. *My Posse Don't Do Homework*. New York: St. Martin's.

Johnson, Richard. 1980. "Cultural Studies and Educational Practice." *Screen Education* 34: 5–16.

Jones, Augustus J., Jr. 1992. "Kindler, Gentler? George Bush and Civil Rights." In *Leadership and the Bush Presidency: Prudence or Drift in an Era of Change?* Edited by Ryan J. Barilleaux, and Mary E. Stuckey, 177–190. Westport, CT: Praeger.

Jorgensen, Joseph G. 1986. "Federal Policies, American Indian Polities and the 'New Federalism.' " *American Indian Culture and Research Journal* 10 (2): 1–13.

Jubera, Drew, and Rosalind Bentley. 2005. "Klan Killer Gets the Maximum." *Atlanta Journal-Constitution*, home edition, sec. A, June 24.

Judis, John B., and Ruy Teixeira. 2002. *The Emerging Democratic Majority*. New York: Lisa Drew/Scribner.

Kaplan, Dave. 1986. "Reagan Vetoes Bill to Renew Indian Programs." *Congressional Quarterly Weekly Report* 44: 2366.

Kaplan, George R. 1992. *Images of Education: The Mass Media's Version of America's Schools*. Washington: National School Public Relations Association and Institute for Educational Leadership.

Kehr, Dave. 1989. "*Dead Poets Society* Exists Only for Itself." *Chicago Tribune*, North sports final edition, June 9.

Kellner, Douglas. 1990. *Television and the Crisis of Democracy*. Boulder, CO: Westview.

Kellner, Douglas. 1994. "Cultural Studies and the Gulf War: A Multiperspectival Approach." In *Styles of Cultural Activism: From Theory and Pedagogy to Women, Indians, and Communism*. Edited by Philip Goldstein, 103–129. Newark/London and Toronto: University of Delaware Press/Associated University Presses.

———. 1995. *Media Culture: Cultural Studies, Identity and Politics between the Modern and the Postmodern*. London and New York: Routledge.

Kelly, Michael. 1992. "Though Advisers Differ, Clinton's in Tune with All." *New York Times*, September 13.

Keroes, Jo. 1999. *Tales Out of School: Gender, Longing, and the Teacher in Fiction and Film*. Carbondale and Edwardsville: Southern Illinois University Press.

Kirp, David L. 1989. "Education: The Movie." *Mother Jones*, January, 36–45.

Koch, Cynthia. 1998. "The Contest for American Culture: A Leadership Case Study on the NEA and NEH Funding Crisis." *Public Talk* 1 (2). http://www.upenn.edu/pnc/ptkoch.html.

Kroll, Jack, with Jennifer Foote. 1988. "To Señor, with Love, Brains, and *Ganas*." *Newsweek*, March 14, 62.

Kwiatkowski, Jane. 1999. "All That Brashness is Real," *Buffalo (NY) News*, final edition, sec. F, October 31.

Ladd, Everett Carll. 1993. "Cause for Unease over Clinton's Flip-Flops." *Christian Science Monitor*, April 16, 18.

LaFraniere, Sharon. 1991. "U.S. Has Most Prisoners Per Capita in the World." *Washington Post*, sec. A, January 5.

Lawson, Bill E., ed. 1992. *The Underclass Question*. Philadelphia: Temple University Press.

Lindsay, Vachel. 1963. *Selected Poems*. Edited by Mark Harris. Repr., New York: Macmillan, 1967.

Lott, Tommy. 1992. "Marooned in America: Black Urban Youth Culture and Social Pathology." In *The Underclass Question*. Edited by Bill E. Lawson, 71–89. Philadelphia: Temple University Press.

Lowry, Brian. 1994. Review of *Renaissance Man*. *Variety*, May 30, 51.

Lugg, Catherine A. 1996. *For God and Country: Conservatism and American School Policy*. New York: Peter Lang.

Lumenick, Lou. 1987. "A Weird One from Hollywood's Trash Heap." Review of *The Principal*. *Bergen (N.J.) Record*, sec. D, September 18.

Lusane, Clarence, with Dennis Desmond. 1991. *Pipe Dream Blues: Racism and the War on Drugs*. Boston: South End.

McCarthy, Todd. 1995. " 'Dangerous' Urban Tale Lacks the Proper Punch." Review of *Dangerous Minds*. *Variety*, August 14, 55, 59.

McIntyre, Alice. 1997. *Making Meaning of Whiteness: Exploring Racial Identity with White Teachers*. Albany: State University of New York Press.

Malanowski, Jamie. 1999. "Shifting from Blood and Guts to Heart and Brains." *New York Times*, late edition-final, October 24.

Malecha, Gary Lee, and Daniel J. Reagan. 1992. "George Bush and Congress: The Question of Leadership." In *Leadership and the Bush Presidency: Prudence or Drift in an Era of Change?* Edited by Ryan J. Barilleaux, and Mary E. Stuckey, 59–80. Westport, CT: Praeger.

Maraniss, David. 1995. *First in His Class: A Biography of Bill Clinton*. New York: Simon and Schuster.

Marcazzo, Cono Robert. 1990. Review of *Lean on Me*. *Magill's Cinema Annual 1990*. Pasadena, CA and Englewood Cliffs, NJ: Salem.

Maslin, Janet. 1994. Review of *Renaissance Man*. *New York Times*, sec. C, June 3.

Meeropol, Michael. 1998. *Surrender: How the Clinton Administration Completed the Reagan Revolution*. Ann Arbor: University of Michigan Press.

Meier, Deborah. 2000. "Fiddling with Facts." *Brill's Content*, February, 63–64.

Miller, James Nathan. 1984. "Ronald Reagan and the Techniques of Deception." *Atlantic Monthly*, February, 62–68.

Miller, Julie A. 1988. "Reagan Picks Hispanic College Chief to Succeed Bennett." *Education Week*, September 7, 34–35.

Moen, Matthew, and Kenneth T. Palmer. 1992. " 'Poppy' and His Conservative Passengers." In *Leadership and the Bush Presidency: Prudence or Drift in an Era of Change?* Edited by Ryan J. Barilleaux, and Mary E. Stuckey, 133–146. Westport, CT: Praeger.

Mohr, Ian. 2005. "Band of Brothers: Part 2," *Daily Variety*, March 30, 1.

Montreal Gazette. 1999. "Movie Didn't Get It Right, School Says." November 1.

Moore, Ben. 1991. "Media Education." In *The Media Studies Book: A Guide for Teachers*. Edited by David Lusted, 171–190. London and New York: Routledge.

Moore, W. John. 1995. "Targeting Harlem, not Hollywood." *National Journal*, February 11, 388.

Morris, C. Patrick. 1994. "Indian Self-Determination and the Tribal College Movement: A Good Idea That Not Even Government Can Kill." In *American Indian Policy: Self-Governance and Economic Development*. Edited by Lyman H. Legters, and Fremont J. Lyden, 71–90. Prepared under the auspices of the Policy Studies Organization. Westport, CT: Greenwood.

Morrison, Toni. 1992. *Playing in the Dark: Whiteness and the Literary Imagination*. Repr., Cambridge, MA: Harvard University Press, 1993.

———, ed. 1992. *Race-ing Justice, En-gendering Power: Essays on Anita Hill, Clarence Thomas, and the Construction of Social Reality*. New York: Pantheon Books.

Mullins, Kerry, and Aaron Wildavsky. 1992. "The Procedural Presidency of George Bush." *Political Science Quarterly* 107: 31–62.

Murphy, John M. 2002. "Cunning, Rhetoric, and the Presidency of William Jefferson Clinton." In *The Presidency and Rhetorical Leadership*. Edited by Leroy G. Dorsey, 231–251. College Station, TX: Texas A&M University Press.

Neely, Anthony. 1984. "Government Role in Rooting Out, Remedying Discrimination Is Shifting." *National Journal*, September 22, 1772–1775.

Noddings, Nel. 2005. *The Challenge to Care in Schools: An Alternative Approach to Education*. 2nd ed. New York: Teachers College Press.

Noriega, Chon A. 1988–1990. "Chicano Cinema and the Horizon of Expectations: A Discursive Analysis of Film Reviews in the Mainstream, Alternative, and Hispanic Press, 1987–1988." *Aztlán: A Journal of Chicano Studies* 19 (2): 1–32.

O'Brien, Tom. 1989. "School of Hard Knocks: Avildsen's *Lean on Me*." Review. *Commonweal*, April 21, 245.

O'Reilly, Kenneth. 1995. *Nixon's Piano: Presidents and Racial Politics from Washington to Clinton*. New York: Free Press.

Omi, Michael, and Howard Winant. 1986. *Racial Formation in the United States: From the 1960s to the 1990s*. Repr., New York and London: Routledge & Kegan Paul, 1994.

———. 1993. "The Los Angeles 'Race Riot' and Contemporary U.S. Politics." In *Reading Rodney King/Reading Urban Uprising*. Edited by Robert Gooding Williams, 97–114. New York and London: Routledge.

Orman, John. 1987. *Comparing Presidential Behavior: Carter, Reagan, and the Macho Presidential Style*. Westport, CT: Greenwood.

Parry, Robert. 2000. "CIA Admits Tolerating Contra-Cocaine Trafficking in 1980s." Consortiumnews.com, June 8. http://www.consortiumnews.com/2000/060800a.html

Perkins, Joseph. 1988. "Boom Time for Black America." *Policy Review*, 45: 26–28.

Pettit, Arthur G. 1980. *Images of the Mexican American in Fiction and Film*. College Station, TX: Texas A&M University Press.

Pitts, Greenfield. 1990. "Mr. Keating from *Dead Poets Society*: From Reel to Real." *Education Digest*, December, 3–6.

Procter, David E., and Kurt Ritter. 1996. "Inaugurating the Clinton Presidency: Regenerative Rhetoric and the American Community." In *The Clinton Presidency: Images, Issues, and Communication Strategies*. Edited by Robert E. Denton, Jr., and Rachel L. Holloway, 1–16. Westport, CT: Praeger.

Quirk, Paul J., and Joseph Hinchliffe. 1996. "Domestic Policy: The Trials of a Centrist Democrat." In *The Clinton Presidency: First Appraisals*. Edited by Colin Campbell, and Bert A. Rockman, 262–289. Chatham House, NJ: Chatham House.

Quirk, Paul J., and William Cunion. 2000. "Clinton's Domestic Policy: The Lessons of a 'New Democrat.' " In *The Clinton Legacy*. Edited by Colin Campbell, and Bert A. Rockman, 200–225. New York: Chatham House.

Reeves, Jimmie L., and Richard Campbell. 1994. *Cracked Coverage: Television News, the Anti-Cocaine Crusade, and the Reagan Legacy*. Durham, NC and London: Duke University Press.

Reich, Robert B. 1999. "We Are All Third Wayers Now." *American Prospect*, March/April, 46–51.

Rhines, Jesse Algeron. 1996. *Black Film/White Money*. New Brunswick, NJ: Rutgers University Press.

Riche, Martha Farnsworth. 1994. "The Bean Count Is In! A Promise Fulfilled: Clinton's Appointees Really Are as Diverse as America." *Washington Post*, final edition, sec. C, January 23.

Riley, Richard W. 2002. "Education Reform through Standards and Partnerships, 1993–2000." *Phi Delta Kappan* 83: 700–707.

Roberts, Steven V., Jeannye Thornton, Ted Gest, Matthew Cooper, Robin M. Bennefield, Katia Hetter, Jennifer Seter, Scott Minerbrook, and Mike Tharp. 1995. "Affirmative Action on the Edge." *U.S. News and World Report*, February 13, 32–38.

Rockman, Bert A. 2000. "Cutting with the Grain: Is There a Clinton Leadership Legacy?" In *The Clinton Legacy*. Edited by Colin Campbell, and Bert A. Rockman, 274–294. New York: Chatham House.

Rogin, Michael Paul. 1987. *"Ronald Reagan," the Movie: And Other Episodes in Political Demonology*. Berkeley and Los Angeles: University of California Press.

Rowan, Cart T. 1991. *Breaking Barriers: A Memoir*. Boston, Toronto: Little, Brown.

Roy, Yancey. 1997. "Nation's Top Educator Praises New York." Gannett News Service, April 30.

Ruffin, David C. 1987. "Picking Federal Judges: Color-blind or Blind to Blacks." *Black Enterprise*, April 25.

Saillant, Catherine. 1995. "School of Soft Knocks; Rio Real Principal, Teachers Reach Out to Parents in Casual Visits to Their Homes." *Los Angeles Times*, sec. B, October 11.

Sanello, Frank. 1984. "*Teachers* Box-Office Champ Three Weekends in a Row." United Press International.

Scalzi, John. 1995. "*My Posse Don't Do Homework* Educator Sees Her Life on Big Screen." *Fresno Bee*, sec. E, August 17.

Schlesinger, Arthur M., Jr. 1991. *The Disuniting of America*. Knoxville, TN: Whittle Direct Books.

Schneider, William. 1992. "A Loud Vote for Change." *National Journal*, November 7, 2542–2544.

Scott, Matthew. 1988. "The Hot Debate: Should Joe Clark Carry a Big Stick?" *Black Enterprise*, May, 38.

Seidenberg, Robert. 1989. "*Dead Poets Society*: Reciting Tom Schulman." *American Film*, July–August, 57.

Shapiro, Walter, and Diane Camper. 1983. "Reagan's Civil-Rights Woes." *Newsweek*, June 6, 38.

Shull, Steven A. 1989. *The President and Civil Rights Policy: Leadership and Change*. Westport, CT: Greenwood.

———. 1993. *A Kinder, Gentler Racism? The Reagan-Bush Civil Rights Legacy*. Armonk, NY: M. E. Sharpe.

———. 1999. *American Civil Rights Policy from Truman to Clinton: The Role of Presidential Leadership*. Armonk, NY: M. E. Sharpe.

Siegel, Frederick F. 1997. *The Future Once Happened Here: New York, D.C., L.A., and the Fate of America's Big Cities*. New York: Free Press.

Silverman, Debora. 1986. *Selling Culture: Bloomingdale's, Diana Vreeland, and the New Aristocracy of Taste in Reagan's America*. New York: Pantheon Books.

Simmons, Althea T. L. 1984. "Civil Rights and Reagonomics." *Crisis*, April, 8–10.

Siskel, Gene. 1989. "*Dead Poets Society* Is Refresher Course in Drama." Review. *Chicago Tribune*, sec. A, June 9.

Smith, Marshall S., Jessica Levin, and Joanne E. Cianci. 1997. "Beyond a Legislative Agenda: Education Policy Approaches of the Clinton Administration." *Educational Policy* 11: 209–226.

Snead, James. 1994. *White Screens, Black Images: Hollywood from the Dark Side*. London and New York: Routledge.

Solowey, Barry. 2000. "Mr. Solowey Speaks Out." *American Recorder* 41 (1): 11.

Stallings, D. T. 2002. "A Brief History of the U.S. Department of Education, 1979–2002." *Phi Delta Kappan* 83: 677–683.

Stein, Ruthe. 1999. "Taking a Stab at Drama; *Scream* Creator Craven Directs Streep in Truth-based Story of Courageous Teacher," *San Francisco Chronicle*, October 24.

Stemer, Rosalie. 1995. "Former Principal Depicted in *Lean on Me* Takes on a Tougher Role." *New York Times*, sec. B, August 14.

Sterritt, David. 1995. "Hollywood Founders Where Issues of Race Are Concerned." *Christian Science Monitor*, November 16, 11.

Stewart, James B. 1990. "Back to Basics: The Significance of Du Bois's and Frazier's Contributions for Contemporary Research on Black Families." In *Black Families: Interdisciplinary Perspectives.* Edited by Harold E. Cheatham, and James B. Stewart, 5–27. New Brunswick, NJ: Transaction.

Sussman, Michael H. 1984. "Reagan's Record of Civil Rights Retrenchment." *Crisis*, April, 28–31, 34–37.

Taubin, Amy. 1995. "Girls in the Hood." Review of *Dangerous Minds. Village Voice*, August 22, 52.

Taylor, Clyde. 1991. "The Re-Birth of the Aesthetic in Cinema." *Wide Angle* 13 (3–4): 12–30.

Thompson, Kristin. 1988. *Breaking the Glass Armor: Neoformalist Film Analysis.* Princeton: Princeton University Press.

Thompson, Robert J., and Carmine Scavo. 1992. "The Home Front: Domestic Policy in the Bush Years." In *Leadership and the Bush Presidency: Prudence or Drift in an Era of Change?* Edited by Ryan J. Barilleaux, and Mary E. Stuckey, 149–164. Westport, CT: Praeger.

Trend, David. 1994. "Nationalities, Pedagogies, and Media." In *Between Borders: Pedagogy and the Politics of Cultural Studies.* Edited by Henry A. Giroux, and Peter McLaren, 225–241. London and New York: Routledge.

Trier, James. 2002. "Exploring the Concept of 'Habitus' with Preservice Teachers through the Use of Popular School Films." *Interchange* 33 (3): 237–260.

———. 2003. "School Film 'Videocompilations' as Pedagogical Texts in Preservice Education." *Journal of Curriculum Theorizing* 19 (1): 125–147.

Tunney, Tom. 1994. Film Review of *Renaissance Man. Sight and Sound*, August, 50–51.

United Press International (UPI). 1989. "Clark on Inner City Schools: Doomed and Rotten to the Core." April 21.

U.S. Bureau of the Census. 1995. *Statistical Abstract of the United States: 1995.* Washington, DC: U.S. Government Printing Office.

———. 2000. *Population Profile of the United States: 2000.* Washington, DC: U.S. Government Printing Office.

U.S. Department of Education, National Center for Education Statistics. 1996. *Digest of Education Statistics 1996.* Washington, DC: U.S. Government Printing Office.

U.S. Department of Labor, Office of Policy Planning and Research. 1965. *The Negro Family: The Case for National Action.* Washington, DC: U.S. Government Printing Office.

van Dijk, Teun A. 1998. *Ideology: A Multidisciplinary Approach*. London: Sage.

Walters, Ronald W. 2003. *White Nationalism, Black Interests: Conservative Public Policy and the Black Community*. Detroit: Wayne State University Press.

Walton, Hanes, Jr. 1997. *African American Power and Politics: The Political Context Variable*. New York: Columbia University Press.

Walton, Hanes, Jr. and Robert C. Smith. 2000. *American Politics and the African American Quest for Universal Freedom*. New York: Longman.

Watson, Peter. 2001. *The Modern Mind: An Intellectual History of the 20th Century*. New York: HarperCollins.

Williams, Linda F. 1998. "Race and the Politics of Social Policy." In The *Social Divide: Political Parties and the Future of Activist Government*. Edited by Margaret Weir, 417–463. Washington, DC: Brookings Institution Press/New York: Sage.

Williams, Raymond. 1979. "Television and Teaching. An Interview with Raymond Williams." *Screen Education* 31 (Summer): 5–14.

Wilmington, Michael. 1989. Film Review of *Dead Poets Society. Los Angeles Times*, sec. 6, June 2.

Wloszczyna, Susan. 1999. "Streep's *Music of the Heart* Plucks All the Right Strings," *USA Today*, final edition, sec. E, October 29.

Wood, Robin. 1986. *Hollywood from Vietnam to Reagan*. New York: Columbia University Press.

Wright, Elizabeth. 2002. "Defaming Garvey Again." *Issues & Views*, July 10. http://www.issues-views.com/index.php/article/2000.

Wright, Lawrence. 1994. "One Drop of Blood." *New Yorker*, June 25, 46–55.

Yeager, Holly. 1996. "President to Take on Prop. 209." *Houston Chronicle*, Three Star edition, sec. A, December 21.

Young, Lola. 1996. *Fear of the Dark: "Race," Gender, and Sexuality in the Cinema*. London and New York: Routledge.

Filmography

Animal House. 1978. Dir. John Landis. Universal Pictures.

Bad Boys. 1995. Dir. Michael Bay. Columbia Pictures.

Beauty and the Beast. 1991. Dir. Gary Trousdale and Kirk Wise. Walt Disney Pictures.

Beverly Hills Cop II. 1987. Dir. Tony Scott. Paramount Pictures.

The Birth of a Nation. 1915. Dir. David W. Griffith. David W. Griffith Corp.

Blackboard Jungle. 1955. Dir. Richard Brooks. Metro-Goldwyn-Mayer.
Boyz N the Hood. 1991. Dir. John Singleton. Columbia Pictures.
Class of 1984. 1982. Dir. Mark L. Lester. United Film Distribution Company.
Conrack. 1974. Dir. Martin Ritt. Twentieth Century Fox.
Crimson Tide. 1995. Dir. Tony Scott. Buena Vista Pictures.
Crying Game. 1992. Dir. Neil Jordan. Miramax Pictures with Channel Four Films (UK).
Dangerous Minds. 1995. Dir. John N. Smith. Hollywood Pictures.
Dead Poets Society. 1989. Dir. Peter Weir. Touchstone Pictures.
The English Patient. 1996. Dir. Anthony Minghella. Miramax Films.
Fast Times at Ridgemont High. 1982. Dir. Amy Heckerling. Universal Pictures.
First Blood. 1982. Dir. Ted Kotcheff. Orion Pictures.
Forrest Gump. 1994. Dir. Robert Zemeckis. Paramount Pictures.
The George McKenna Story (television). 1986. Dir. Eric Laneuville. Landsburg Company.
Ghost. 1990. Dir. Jerry Zucker. Paramount Pictures.
Glory. 1989. Dir. Edward Zwick. TriStar Pictures.
Good Morning, Vietnam. 1987. Dir. Barry Levinson. Touchstone Pictures.
Good Will Hunting. 1997. Dir. Gus Van Sant. Miramax Films.
Goodbye, Mr. Chips. 1939. Dir. Sam Wood. Metro-Goldwyn-Mayer.
The Graduate. 1967. Dir. Mike Nichols. Embassy Pictures Corporation.
Grease. 1978. Dir. Randal Kleiser. Paramount Pictures.
Iron Eagle. 1986. Dir. Sidney J. Furie. TriStar Pictures.
Iron Eagle II. 1988. Dir. Sidney J. Furie. TriStar Pictures.
The Journey of August King. 1995. Dir. John Duigan. Miramax Films.
The Karate Kid. 1984. Dir. John Avildsen. Columbia Pictures.
The Last House on the Left. 1972. Dir. Wes Craven. Hallmark Releasing.
Lean on Me. 1989. Dir. John G. Avildsen. Warner Bros.
Mr. Holland's Opus. 1995. Dir. Stephen Herek. Hollywood Pictures.
Music of the Heart. 1999. Dir. Wes Craven. Miramax Films.
Navy Seals. 1990. Dir. Lewis Teague. Orion Pictures.
A Nightmare on Elm Street. 1984. Dir. Wes Craven. New Line Cinema.
The Paper Chase. 1973. Dir. James Bridges. Twentieth Century Fox.
Pretty Woman. 1990. Dir. Garry Marshall. Touchstone Pictures.
The Principal. 1987. Dir. Christopher Cain. TriStar Pictures.
The Program. 1993. Dir. David S. Ward. Touchstone Pictures.
Pulp Fiction. 1994. Dir. Quentin Tarantino. Miramax Films.
Rambo: First Blood Part II. 1985. Dir. George P. Cosmatos. TriStar Pictures.

Renaissance Man. 1994. Dir. Penny Marshall. Touchstone Pictures.
Rising Sun. 1993. Dir. Philip Kaufman. Twentieth Century Fox.
Rocky. 1976. Dir. John Avildsen. United Artists.
The Scarlet Letter. 1995. Dir. Roland Joffé. Hollywood Pictures.
Scream. 1996. Dir. Wes Craven. Dimension Films.
Scream 2. 1997. Dir. Wes Craven. Dimension Films.
Shakespeare in Love. 1998. Dir. John Madden. Miramax Films.
Small Wonders. 1995. Dir. Allan Miller and Lana Miller. Four Oaks Foundation. Released 1996. Miramax Films.
Sorority House Massacre. 1987. Dir. Carol Frank. Concorde Pictures.
Stand and Deliver. 1988. Dir. Ramón Menéndez. Warner Bros.
Teachers. 1984. Dir. Arthur Hiller. Metro-Goldwyn-Mayer.
Terminator 2: Judgment Day. 1991. Dir. James Cameron. TriStar.
To Sir, with Love. 1967. Dir. James Clavell. Columbia Pictures. Original release in UK, 1966, Columbia Pictures Corporation Ltd.
Top Gun. 1986. Dir. Tony Scott. Paramount Pictures.
Up the Down Staircase. 1967. Dir. Robert Mulligan. Warner Bros.
The Wizard of Oz. 1939. Dir. Victor Fleming and King Vidor. Metro-Goldwyn-Mayer.

Index